Incident Management for the Street-Smart Fire Officer

Incident Management for the Street-Smart Fire Officer

John (Skip) Coleman

FIRE ENGINEERING®

Disclaimer

The recommendations, advice, descriptions, and methods in this book are presented solely for educational purposes. The author and publisher assume no liability whatsoever for any loss or damage that results from the use of any of the material in this book. Use of the material in this book is solely at the risk of the user.

Copyright © 1997 by Fire Engineering Books & Videos,
a Division of PennWell Publishing Company.

Published by Fire Engineering Books & Videos
A Division of PennWell Publishing Company
Park 80 West, Plaza 2
Saddle Brook, NJ 07663
United States of America

EDITED BY MARY JANE DITTMAR
BOOK DESIGN BY JULIE HENDRICKSON
COVER DESIGN BY CLIFF RUMPF
PHOTOS BY GERRY MAZUR, EXCEPT AS NOTED

2 3 4 5 6 7 8 9 10

Printed in the United States of America

Library of Congress Cataloging-in-Publication Data
Coleman, John, 1950-
Incident management for the street-smart fire officer/John (Skip) Coleman.
p. cm.
Includes index.
ISBN 0-912212-60-8 (hardcover)
1. Fire extinction. 2. Emergency management.   I. Title.
TH9310.5.C636   1997
363.37′368—dc21

                                                          96-46764
                                                          CIP

About the Author

John (Skip) Coleman has been a member of the Toledo (OH) Department of Fire and Rescue Operations for 21 years and has been a battalion chief for the past 11 years.

Coleman has been an instructor at Owens Community College, one of Ohio's largest community colleges, for more than 10 years.

Additionally, he is a contract instructor for the National Fire Academy's Command and Control of Fire Department Operations at Multialarm Incidents course and annually conducts a course in incident command for industrial fire brigades at the Ohio State Fire School.

He is a graduate of the National Fire Academy's Executive Fire Officer Program and is working toward his bachelor's degree at the University of Cincinnati.

He has been married to Theresa Ann for 17 years. They have two children, Toby and Betsy; a dog, Chessie; and Betsy's horse, Bailey.

To Theresa, my wife.
You make everything in my life possible
and worthwhile!

Acknowledgments

Many people have made significant contributions to this text—some by providing services or information; others through the influence they have had merely by being present in my life. I thank and personally acknowledge the following individuals:

My parents. My father, for teaching me how to golf and fight fires. My mother, for teaching me right from wrong and that patience is a virtue. Both, for instilling in me the belief that I can do just about anything I put my mind to.

Deputy Chief Robert Schwanzl, who in 1984 had enough faith in me to ask me to come to the Training Bureau to help teach a fire recruit class. (That experience, in addition to my parents' influence, represents the true beginning of this book.)

Battalion Chiefs Joe Walter and Mike Wolever (the other Platoon A chiefs), who helped me keep my sanity and sense of humor while on duty.

All members of Platoon A, who make work fun and a lot easier than it looks.

Chief Alan V. Brunacini, Chief Ronny Coleman, Chief Neil Honeycutt, and Chief William Goldfeder, for their contributions to this text in the form of inspiration and information.

Gerry Mazur, for the photos.

Mary Jane Dittmar, for her editorial assistance.

Bill Manning, who believed that this text was worth publishing.

I know that I have failed to mention someone who absolutely should be acknowledged. To that individual, I simply say, "Thank you."

The author and publisher recognize the significant contributions female firefighters, fire officers, and chief officers have made and are making to the fire service. The gender references used in this text encompass both female and male.

Table of Contents

Section I - The System

Section II - Staging and Sectorization

Section III - Mission Statements

Section IV - Miscellaneous Incidents

Section V - Applying Incident Command

Preface

Incident management as described in this book is not a panacea. The methods I espouse "fit what I do." I have never been to a wildland fire bigger than a few football fields. My department does not own a helicopter. I have a fire hydrant at least every 600 feet of anywhere within my jurisdiction with only a handful of exceptions. I have more than a hundred firefighters on duty 365 days a year.

I realize that the forestry industry has a working model of the incident management system already in place. It would be foolish of me to try to write a text on an incident management system that could be used in wildland firefighting. Many fire departments, large and small, paid and volunteer, use a "known" form of the IMS. Other departments use parts or combinations of "known" systems. (As I am writing this, I am confident that there has to be at least one department out there with its own system. That system, if compared with the known versions, would undoubtedly share many similarities.)

This text is directed at any department, large or small, that does not use some form of incident management, or which knows deep down that the one it has may look good on paper but isn't practicable on the fireground.

I have attempted to show how a practical working model of incident management can be used in any fire department, on any type of incident. Many books and pamphlets previously written on the subject have used vast flowcharts indicating dozens of sectors and strike teams. These systems work—and work well. This type of system is viable when numerous teams of personnel with vast resources can be gathered to fight huge fires. What I have always felt these books lack, however, is the applicability of such a system to everyday incidents such as a car fire.

Incident management works. It works in California, Pittsburgh, Toledo, and Phoenix. Its applicability to everyday incidents is the key to its success. I have used it at every incident to which I have responded since 1988. (My department responded to more than 50,000 incidents in 1996 alone.) I wouldn't even consider going to an incident where incident command isn't used.

The use of the incident management system is now mandatory for certain kinds of incidents, regardless of jurisdiction. In Ohio, recent legislation has made it a state law that the IMS be used by all state and state-funded agencies. That includes prisons, police departments, and fire departments receiving state monies.

If your department currently uses the IMS, this text may provide some new twists to the system. It may explain why you need to do some of the things you are doing. I have included some new concepts that have been proved, such as mission statements for all working sectors at a fire or emergency medical service incident. I have figured out a way to maintain sections such as Logistics and Liaison even at single-family structure fires. I have included the police in some sectors at incidents in which they also have concerns. I have also introduced some relatively new concepts such as rapid intervention teams and positive-pressure ventilation.

Additionally, this text meets and exceeds the minimum requirements of NFPA Standard 1561, *Fire Department Incident Management System,* and complies with the *Model Procedures Guide for Structural Firefighting* as established by the National Fire Service Incident Management System Consortium.

This text will be primarily concerned with use of the incident management system at incidents that can be handled by five or fewer companies. The intent is to show how the system can work at our bread and butter incidents, not to "wow" you with overwhelming scenarios.

Section I

The System

Introduction to the Incident Management System

TOPICS

- The history of the incident command system (ICS)

- The "basic" differences between the Phoenix (AZ) Fire Department and the National Fire Academy (NFA) forms of incident command

---

In the mid 1980s, I experienced what up to now has been the largest fire of my career. It involved a warehouse property of approximately six acres bordering on a residential area. The fire started on a summer weekend evening. At that time, I was a captain in the Toledo (OH) Department of Fire & Rescue Operations (for brevity and nostalgia, often referred to in this text as the "Toledo Fire Department," its previous name) and was dispatched to the fire on the second alarm as the third-due engine. That was around 2000 hours. Within two hours, that fire beat us. It used up our resources of some 110 on-duty firefighters, 17 engine com-

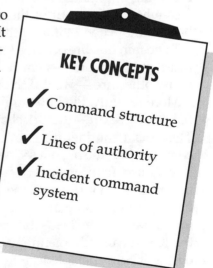

KEY CONCEPTS

✓ Command structure

✓ Lines of authority

✓ Incident command system

panies, five truck companies, and two heavy squads and left us unable to advance. We were released at shift change at 0700 the next morning. I spent 11 hours at this fire.

I still remember many aspects of that fire: chiefs circling the fire; conflicting orders; and, during one period of time, a chief on the roof punching ventilation holes with an ax. I also remember going through eight self-contained

Figure 1-1. Within two hours, the fire beat us. It used up our resources and left us unable to advance.

breathing apparatus (SCBA) bottles and the look of surprise on everyone's face when we ran out of them at the scene. Above all, what keeps coming back to me when I think of that fire is its lack of coordination.

To be fair, this fire occurred three years before our department adopted the incident command system (ICS). We fought that fire the way we'd fought many other fires—we did the best we knew how to do.

More recently, the Pittsburgh (PA) Bureau of Fire responded to a residential structure fire—the Bricelyn Street fire—in which three firefighters died. This was a routine fire—nothing like the Toledo industrial fire. We expect confusion at "the big ones." We expect control and coordination at the everyday house fires, what I refer to as our "bread and butter" fires. What occurred at this fire and the Toledo fire basically was the same thing—a breakdown in command (the individual in charge of the fire). At the Toledo industrial fire, the breakdown didn't occur because we had failed to follow the ICS. As already mentioned, that fire occurred before we had implemented the ICS in the Toledo Fire Department. That fire lacked a "centralized" command figure who could stay back and look at what was happening and anticipate strategies and equipment needs on the larger scale, as opposed to climbing ladders and circling the structure a hundred times.

Pittsburgh, on the other hand, did have an ICS in place prior to the fire. According to the reports I received, however, it broke down that night.

Two separate fires, in different decades, in different cities—yet both had unsuccessful outcomes due to the same cause: the lack of a centralized command system.

As a result of the Toledo incident, our department made several changes in our response and operations at large industrial complex fires. Although not immediately, this fire led to the introduction of the incident command system in the Toledo Fire Department. I know that implementing this sytem has enhanced our operations.

After the Bricelyn Street fire, the Pittsburgh Bureau of Fire conducted an extensive investigation and, according to Chief Charlie Dickenson, "did a lot of soul-searching." The end result to this date[1] has been the implementation of several changes in the department's operations, including the following:

- A stronger focus has been placed on the incident command system.
- Ten support people are now sent to every working fire. Some of the tasks performed by support personnel include the following:
  —A chief officer takes the role of safety officer.
  —A specific engine company assists Command by setting up the "command board" and filling other "staff" positions.
  —A specific company serves as a rapid intervention team (RIT).

A UNIFORM COMMAND SYSTEM IS A NECESSITY

I wrote this text for two reasons. The first is that I want to have an up-to-date text on incident command that I can use when I teach. The most recently published texts on incident command do not even list it in the index. Some address the "whats" of the system but not the "hows" and "whys," or they cover the system only as it relates to firefighting operations. As we know, only a small portion of our day is filled with firefighting. Later, we will see how incident command can be adapted for all emergency incidents—fire, haz-mat, and medical.

The second reason I wrote this book is to support and perhaps boost the idea of an "identifiable" uniform ICS. There are almost as many systems out there as departments. I have been conducting surveys related to incident command for my department on and off since 1988. The most recent was in March 1996. I found that in general most departments in the country basically know what the ICS is, most have a procedure for using it when the incident gets "big enough," and there is no "standard" system in use.

During the last survey, I conducted a phone interview with an operations chief of one of the country's 10 largest departments. The first question I asked him was, "Do you use the incident command system?" He said, "Yes."

I then asked him, "Would you use the ICS at a single-family residential structure fire?"

He answered, "We don't do that @&%$^. We have trained officers. We use specific procedures at structure fires, two engines, and two trucks." He went on to list the tasks performed by the engines, the first truck, and the second truck. He closed with, "We train our officers well. They know what needs to be done."

Because I was conducting a survey and not an inquisition, I did not ask him any questions other than those on the survey.

But, come on! Procedures? According to that chief, the department has a "single-family residential structure fire procedure." Let's say that this procedure is that the first engine takes a line in to attack the fire, the second engine provides water to the first engine and then goes above the fire to check for extension, the first truck vents the roof, and the second truck ladders the rear of the structure. What happens if people are hanging out of the front windows when the responders pull up and heavy volumes of fire are in the rear of the basement? Where is "search" in the procedure, and what is the priority of a search in this scenario? Are we still going to ladder the building in the rear if the fire is in the rear? Where is the backup line? If the answer is, they know what needs to be done, then who determines who does it and when?

I am a firm believer in procedures. There must be procedures that explain what to do if valuables are found at a fire scene or if apparatus break down on the street. And there should be emergency procedures for disasters such as floods and tornadoes.

When we went to the ICS at *all* incidents, we threw out most of our structural firefighting procedures. In my opinion, unless you are going to have a procedures manual filled with *every* eventuality, you don't need specific firefighting procedures. Don't confuse procedure with evolutions. We need training manuals that define hose evolutions—pull and stretch 2½-inch preconnected attack lines—and ladder evolutions such as the four-man 35-foot right-angle raise, and so on.

The ICS provides for "variations" at every incident because every incident has specific, unique variations. Incident command demands that one person determine—not procedures dictate—what needs to be done, when, and by whom.

TRADITION IN THE FIRE SERVICE

As the saying goes, "The fire service is 150 years of tradition unhampered by progress." Change is a slow, painful process in the fire service, and new advances must be proved to be accepted. As a recruit firefighter in 1976, I was assigned to #6 station on the east side of Toledo. At 2300 hours, it was procedure for the night watchman to wash the wheels of the rigs. Every night—summer or winter—you washed the wheels at 2300 hours if you had night watch. A group of us younger firefighters finally gathered the courage to ask, "Why?" The answer was in some ways shocking, but if you looked at it from all sides (keeping in mind the first sentence of this section), it made sense. We washed the wheels to clean off any horse droppings. Now, bear in

Figure 1-2. The fire service is 150 years of tradition unhampered by progress. Shown is the Toledo Fire Department's #6 engine on March 2, 1904. (Courtesy of the Toledo Fire Museum.)

mind that the last time a horse-drawn fire engine was used in Toledo was during World War I. Nevertheless, tradition had it that at 2300 hours, the watchman washed the wheels. Sixty years after the horses were gone, we were still washing wheels. Such is the resistance that innovation in the fire service is up against.

HISTORY OF INCIDENT COMMAND

FIRESCOPE

In California, during the late 1960s and early 1970s, frustrations were running high after a rash of several major wildland fires. It was evident that there was little coordination between municipal, county, and state agencies at the larger fires. In 1970, a particularly large wildland fire was burning out of control. Resources were being stretched, and the fire was spanning over several jurisdictions. That there was an abundance of confusion and a lack of control became evident after the fire (and tempers) cooled down. As all fires do, the fire eventually went out.

After this particular fire, however, the chief officers involved sat down and decided that there had to be a better way of figuring out who was doing what—and where—at fires and who had the final say at large incidents. The incident command system was born out of these meetings. Federal money was provided so that a statewide system to handle these large fires could be adopted. The ICS basically was patterned after the system used by the military to control its forces in battle. One person is put in charge. Prior to the advance, assignments are given and expectations are defined. Terms such as "sector," "logistics," and "support" are used.

The system developed in California was called *FIRESCOPE* (Firefighting Resources of Southern California Organized for Potential Emergencies). It is still in use today. It features a specific hierarchy followed by several layers of support groups from across the state. It emphasizes the roles and responsibilities of all members of the team.

APPLICATION FOR URBAN FIREFIGHTING

Eventually, urban firefighters noticed the ICS. This system could give control and organization to urban as well as rural firefighting. With a few modifications, its use soon spread to the major fire departments on the West Coast. In fact, when I first learned of the ICS in 1987, it was called "the California method" of firefighting. As the use of the ICS spread eastward, it was modified to fit some specific areas. Soon it was introduced to the National Fire Academy (NFA), where it gained acceptance and was added to the curriculum. "Preparing for Incident Command" (PIC) and "Commanding the Initial Response" (CIR) were the two most noteworthy courses offered by the NFA. PIC and CIR became buzzwords in the fire service. The West Coast and the NFA became the advocates of this formal method of operating, which eventually came to be known as *incident command*.[2]

The Phoenix Fire Department

In the early 1980s, Chief Alan Vincent Brunacini of the Phoenix (AZ) Fire Department began looking at the ICS and incorporating its basic premise into his operation. His philosophy toward firefighting and commanding firefighting operations was somewhat less formal and more "laid back" than the formal method (the ICS) used in California. However, he did see that some aspects of the system filled voids that existed in his department and soon developed his own version of the ICS. It was far less structured and geared more toward traditional firefighting.

The Toledo Fire Department

In the spring of 1988, I, as chief of training for the Toledo Fire Department, was asked by Department Chief William Winkle to look at the ICS and make it "fit" the department. That assignment was not as easy as it sounded. In 1988, not many departments operated under the ICS, especially in the Midwest and eastern part of the country, where most of my fire service acquaintances were located. Some departments responding to my questionnaire obviously were using the ICS in name only. Procedures were weak or nonexistent.

After much searching, I found that two basic forms of the ICS were in use: the National Fire Academy's version (basically, the California method or *FIRESCOPE*) and the Phoenix (or Brunacini's) method. These two versions dominated until the early 1990s. While sitting in the "Pub" at the National Fire Academy, you would hear (or participate in) countless discussions about which system was best—which system was the "real" system.

BASIC DIFFERENCES BETWEEN THE NFA AND PHOENIX SYSTEMS

The National Fire Academy's form of the ICS contained a definite command structure but was full of items more suited to wildland firefighting, not urban structural firefighting. Chief Brunacini's version certainly had command structure but omitted the large-scale operation items.

The differences were manifested in the following ways:

- *The responsibilities of the first-arriving officer.* In the NFA method, the first officer was required to establish command and to direct incoming units until relieved by a higher authority. This gave every incident a strong command presence from the time the first unit arrived.

    Under the Phoenix method, the first officer could, under certain circumstances, *pass* his command responsibilities to the next-arriving officer. This would allow more personnel to actually participate in the initial attack or rescue operations.

- *The responsibilities of command.* The NFA form of the ICS was very formal. Flowcharts were used to emphasize the command structure and lines of authority. Basically, command was responsible for four functions at every incident: Operations, Planning, Logistics, and Finance. All activities needed to mitigate a fire were covered under these four functions.

    Chief Brunacini believed it would be difficult at a house fire to tell his officers that they would be responsible for financing the incident. He took a more simplistic approach to the responsibilities of command. Under his system, Command is responsible for firefighter safety, civilian safety, stopping the fire, and conserving property—responsibilities he believed would be much more acceptable to his officers.

- *Method of dividing the incident (sectors).* Some terms in the two forms of the ICS were changed or deleted. Terms such as "sector," "division," and "strike team" had different meanings in the versions or were not present. Additionally, under the NFA's method, more than 260 separate sectors could operate at one incident without exceeding the recommended span of control. The Phoenix method allowed for slightly more than 60 sectors.

One problem that developed from the differences in the two systems was that not all communities in the same mutual-aid pacts were using the same system. This situation led to confusion on the fireground—ironically created by the very system that was supposed to resolve fireground confusion.

THE COMPROMISE/THE FUTURE

The solution to this problem may be at hand. A consortium of firefighting experts, including Chief Brunacini and members of the NFA, have developed a system that may placate both "camps." This new system, called the *incident management system (IMS)*, is the version of the command system discussed in this text from this point on. It incorporates the best features of the Phoenix and NFA methods into what I believe now represents a "working" model of the incident management system that can be applied to all situations.

THE HISTORY OF THE CONSORTIUM

The first vestiges of the incident command system sprang up in the early 1970s. At about that time, while the forestry service in California was working on *FIRESCOPE*, Phoenix was looking at methods to better control and coordinate the actions of the department's crews at normal incidents. *FIRESCOPE* was tailored to serve expansive incidents with hundreds of firefighters and crews and was more formal and structured. Phoenix tailored its new management system to fit up to about a third-alarm incident and was more "how-to" based and less formal.

In the late 1980s, after almost 10 years of bickering, the International Association of Fire Chiefs (IAFC) asked Chief Brunacini and members of the *FIRESCOPE* camp to get together to see if some areas of commonality could be created to "stop the madness." Five or six members of the Phoenix Fire Department met with a similar number of representatives of *FIRESCOPE* in Phoenix in July 1990. These meetings lasted several days and were moderated by a representative of the National Fire Academy. The consortium was derived from those meetings.

Other meetings were scheduled, and areas of commonality were drawn. To be sure, there were still some areas of contention and disagreement—the "s" word (sector) is still a sore spot with some, for example. However, progress was being made, and it was determined that a reference of the common terms and practices would be put together in a text for others to share. In February 1993, the final touches were put on the *Model Procedures Guide for Structural Firefighting*, published by Fire Protection Publications (FPP), which holds the copyright. At the time this book was published, it was expected that guides for high-rise firefighting and mass-casualty incidents would be available by the end of 1996 or early 1997.[3]

Endnotes

1. I say "to this date" because after listening to Chief Dickenson speak, and after discussion with several Pittsburgh chiefs, it is evident that the department understands that minor modifications may be needed.

2. To accommodate urban fire services and other emergency agencies not associated with wildland firefighting, a National Interagency Incident Management System (NIIMS) was created. It had all the features of *FIRESCOPE* but was geared more for "all-risk" incidents. For the most part, the National Fire Academy uses *FIRESCOPE* as its basis for incident management.

3. Thanks to Deputy Chief Gary Morris of the Phoenix (AZ) Fire Department for providing me with the above history of the consortium. Chief Morris is chairman of the National Fire Service Incident Management Consortium.

REVIEW QUESTIONS

1. Describe the incident that brought about the beginning of the incident command system.

2. At what levels was the funding for this project supported?

3. What was the first form of the incident command system (ICS) called? Who used it?

4. What are the two major forms of the incident management system (IMS) currently in use today?

5. List the three basic differences between the two major forms of the IMS.

DISCUSSION QUESTIONS

1. The author notes that "Change is a slow, painful process in the fire service." Discuss the change process in your department. How does the department deal with innovation and change?

2. Write down your department's incident management structure. If your department does not use the IMS, which form of the system do you feel would best fit your department?

3. Discuss the need for a strong command presence at an incident. Give examples of incidents to which you recently responded that you felt needed a stronger command presence.

2

Command

The incident management system (IMS) is a standard method of operating at all incidents to which the fire department responds. It is a management tool that defines the roles and responsibilities of *all* units responding to an incident. It enables one individual to totally control the incident and should eliminate any form of freelancing.

In the "old" days, it was not uncommon for an officer to respond to a structure fire and be allowed to place his apparatus anywhere on the fireground and operate as he saw fit. The fires went out. (Fires always eventually go out.) Also, prior to the IMS, there often was little coordination and control at incidents. It was easy for all at the scene—recruit firefighters to the highest ranking chief officer—to elude responsibility merely by stating, "I didn't know."

KEY CONCEPTS

✔ Formal command

✔ Incident management

✔ Informal command

✔ Macromanaging

✔ Micromanaging

✔ Span of control

✔ Unified command

BASIC PREMISE OF THE IMS

As previously noted, the IMS is as much an *attitude* as a system. When used to its fullest extent, participants feel the operation is controlled, the operation is coordinated, and a team spirit prevails. This attitude surfaces because all the members within the organization know and understand that, for this sytem to work, the following things are needed:

- *Trust among all department members.* This system relies on the principle: "I (Command) will tell you what needs to be done, and you (the sector officer) will take your crew and go do it."
- *An understanding among the crews that the person in charge will be "standing back" from the incident, focusing on the entire scene.* The person in charge will ensure that everything that needs to be done is being done and that someone is watching out for their safety at all times.
- *A basic premise allows the system to work.* Premise components (constituting the focus of the premise) are the following:
  - *One person shall be in charge of every incident to which the department responds.* That person is referred to as *Command*. (When capitalized, "Command" refers to the *individual* in command, as opposed to the incident management component, which is presented in lowercase.)
  - *Command ultimately shall be held responsible for the outcome of the incident.*
  - Most importantly, *if Command is to be held ultimately responsible for the outcome of the incident, that individual must have total control.* Freelancing in any fashion cannot be allowed. Only one individual should set strategies. Only one individual should give assignments.

The IMS enables Command to control any kind of incident response— high-rise, haz-mat, and water-rescue incidents, for example—as long as the above three conditions are adhered to. It doesn't matter if the individual in command is familiar with the jurisdiction, just as long as that individual has total control of "who is doing what—and where." If department members cannot abide by the basic premise behind the IMS, then no matter what they do, what they call it, or how they draw it on a piece of paper, the system won't work, and it won't be the real IMS.

COMMAND

Command is defined as the *person* ultimately responsible for the outcome of the incident—the individual who sets the tone for the incident and defines strategies.

As the basic premise behind the IMS states: *One individual shall be in charge of every incident.* That individual has been referred to by several titles in the past. The NFA and the majority of the departments that adopted the NFA form of incident management generally refer to that individual as "Command." The departments that adopted the Phoenix form of incident management[1] may have referred to the individual in charge as the "Fire Ground Commander (FGC)." It doesn't matter what the individual is called.

What is vitally important is that
- *one—and only one—individual be in charge,*
- everyone responding (or who could respond, as in mutual aid) knows what that individual in charge is called, and
- no one does anything until that person directs him to do it.

WHEN TO USE INCIDENT MANAGEMENT

The use of the IMS should be required at every incident to which the department responds, regardless of the nature of the incident. Several large departments use the IMS only at multiple-alarm incidents. I'm not quite sure how well that works. First, I was taught in my first "Instructional Techniques" course that the human brain forgets 75 percent of what it has learned if it is not reinforced within two weeks. The Toldeo Fire Department is one of the 50 largest departments in the nation. We average only about one multiple alarm a month. With our three-platoon system (24-48), that would mean that each shift gets about four big ones a year. The chance that one of them will be in your district is slight. What are the chances that you will remember all of the nuances of the IMS when it is used only once or twice a year or less frequently? "Am I staging or base? And where the heck is that?" Additionally, who is responsible for the outcome of the typical everyday house fires to which the department responds? This does not mention the "still-box" alarms for single- and double-unit response. I don't see how they can say they have a strong command at these incidents.

Part of the Superfund Amendments and Reauthorization Act (SARA) Title III specifies that any department responding to a hazardous-materials incident shall use the incident command system (aka the IMS). If the department you work for (paid or volunteer) responds to hazardous-materials incidents, even if only at the first-responder level, you fall into this category. If you are dispatched to a haz-mat incident and things go well at this incident, then it's like basketball—"no harm, no foul."

Figure 2-1. Federal law requires the use of the incident management system at all haz-mat incidents.

SCENARIO 1

Let's say that you're part of a 48-man volunteer department in a Midwest state. At 0400 on a Saturday, you are called to respond to I-80 at Exit 7 for a vehicular accident involving a tractor-trailer rig.

The rig, placarded 1203, is jackknifed on the median strip and no visible fire or spill is showing on your arrival. On closer look, you find that the tanker skidded on a patch of ice and jackknifed and that the force of the rig's coming to a quick stop on the median caused one of the saddle tanks to split, dumping about 25 gallons of diesel fuel on the median strip. There is no other spill, and the driver was not injured. A wrecker is called to move the rig. You notify the state environmental protection agency (EPA) of the small spill, and it contracts with a local company to clean the spill site in the morning (the spill was over dirt and soaked into the ground prior to your department's leaving at 0615). You leave, the rig drives away, and one local cleaup contractor is very happy.

But if things don't go well .... Let's say, for example, that the following happens.

SCENARIO 2

When you arrive at the above scene, you notice as you approach that the middle section of the tanker was split by the force of the accident and approximately 2,000 gallons of gasoline are flowing down into a sewer system. A policeman, directing traffic while awaiting a wrecker (smoking a cigarette), accidentally ignites the spilled fuel and a firefighter not in bunker gear is burned. As you attempt to put foam on the flowing fire, you push gasoline, water, and AFFF (aqueous film-forming foam) down the sewer.

The first thing that will happen when the Feds come in (and they will come in) is that they will ask to see your command structure and to talk to whoever was Command. It is better to prepare now by having a strong command structure before a problem occurs.

Another concern is NFPA 1500, *Standard on Fire Department Occupational Safety and Health Program*—1992.[2] I am certainly aware of the components of this standard. The Toledo Fire Department has been struggling with this standard for more than six years now. The standard calls for the IMS as part of the overall scene safety plan at structural fires. The IMS promotes safety, the theme of Chapter 5 and a dominant theme throughout this text.

As stated earlier, the IMS should be used at every incident to which a fire department responds—single-family structure fires, multiple alarms, mutual-aid incidents, emergency medical runs, water rescues, haz-mat incidents, alarm system responses, and so on. It doesn't matter which type of incident it is. If a unit is dispatched, some form of the IMS should be used.

Approximately two years ago, I, an engine company, and the water rescue unit were dispatched to check for a "jumper" on one of our bridges over the Maumee river. The first to arrive, I found a young female over the railing and hanging on to the rail with one arm. She obviously was intoxicated and distraught. I took command and walked to within about 10 feet of her—as close as she would allow me to come. I began to talk to her. When Engine 6 arrived, I quickly told the officer to take command. I couldn't talk to her, provide the attention she needed, and still position the water rescue unit and coordinate with the police. To make a relatively long story short, command was established at this incident. Everyone understood that someone was in charge and that no one would approach the jumper, put divers in the water, or call for a medic unit unless Command directed it. It has to work this way. I've used this system since 1988. I'd never want to go back to what it used to be.

TYPES OF COMMAND

The IMS is extremely flexible. It is designed to work at every type of incident. However, not every incident to which we respond requires the same response level or command structure. Command can be adapted for different response types and sizes.

Some units may roll out the door with more than one officer. Some may not have any officer riding the seat (or anywhere else). If more than one fire department unit responds (other than the normal first responder/medic EMS response), someone has to take charge. Normally, the fire department is the only entity represented at the command post, particularly for most structure fires. Other entities may wander up to the command post looking for someone to talk with, tell their problems to, or bum a cup of coffee from. Although this may be quite common, the fire department still is the only entity responsible for the incident. However, there will be instances when the incident will best be served if the fire department and other entities work together to meet its needs. Haz-mat incidents that involve several jurisdictions, for example, may be best served if the chiefs of the jurisdictions involved jointly share command responsibilities. A large civil disturbance with many fires and looting may require police and fire departments to jointly share the responsibility of command.

The three types of command—informal, formal, and unified—are used in specific situations in accordance with
   • the number of units responding, and
   • the number of entities formally represented at the command post.

Informal Command

Informal command is established when only one officer responds to the incident, such as in normal emergency medical service (EMS) and "still-box" alarm responses. In Toledo, this also includes the normal first responder/medic response on advanced life support (ALS) runs. In these responses, traditionally only one officer responds to the incident on a single unit. (In some rural departments where volunteers are used, more than one officer may respond on these runs. I suggest that procedures designate the highest ranking officer as the informal Command at these incidents.)

Figure 2-2. An engine company operating under informal command. (Photo by author.)

When operating under informal command, command does not have to be formally established or announced. It is understood that only one officer is on the scene and that this officer is Command and, as such, in charge of the incident.

In these responses, when the officer reports on the scene, he should give an initial on-scene report, stating that the unit is on the scene and including an initial conditions report.

Engine 5 is at 1945 Vermont. We have a car totally involved.

At this point, the officer and his crew should handle the incident as training and procedure dictate.

Formal Command

Formal command is used whenever more than one fire unit responds to an incident—for two-engine still-box, one engine-one truck alarm, two-plus-one (two engines and one truck) initial structure responses; vehicular accidents, and the like. Under formal command, the officer assuming command announces that fact over the radio.

The first officer on the scene should give an initial report similar to the one given under informal command but adding which command mode will be used (command modes, which define the initial actions of the crews responding, are discussed in Chapter 3).

Engine 5 is at 1945 Vermont. Smoke showing. Engine 5 is Vermont Command.

Figure 2-3. When multiple units are dispatched to the same incident, formal command is required.

When I taught the chief officers of the Toledo Fire Department the IMS in 1988, I got a lot of "Do you mean to tell me that I have to tell the officers on the scene that I am in charge? I'm the chief and, by God, they ought to know that I'm in charge."

Not too many chiefs can run an incident well while putting on their gear. As you read on, you'll see that I use the term "focus" a lot. If all personnel focus on what they are supposed to be doing, everything gets done well. It's hard to focus on running an incident while you're parking, dressing, or walking up to the incident. You may not even want to take command at the incident. It may be too small an incident for your involvement, or you may want the officer on the promotional list who initially took command to continue to run the fire with you standing nearby so the officer can gain experience and confidence.

Unified Command

Unified command is a newer aspect of the IMS. Used predominantly at large, multijurisdictional incidents, it allows for more than one individual to be Command and establishes a "think tank," if you will, of individuals to share the burden of decision making and responsibility at an incident. An example of a large incident at which unified command would be established would be a large spill on Lake Erie (such as the one simulated in a multi-agency drill in the summer of 1994 in Toledo). A spill of that nature could affect not only the United States but also the Canadian waters of Lake Erie. In this instance, unified command that incorporates representation by Ohio, Michigan, and Canadian fire officials and the United States and Canadian Coast Guard could be established. These five (more or fewer) individuals would jointly make decisions and share responsibility for the outcome of the incident. On a much smaller scale, unified command may be established at wildland fires involving two or more jurisdictions.

When this type of command is used, the individuals jointly filling the role of Command must by some method document that they are all part of a unified command structure. At large incidents (usually of long duration), a "sit-stat" officer,[3] to formally note the individuals jointly acting as Command, probably will be established. Some of the command individuals may take shifts. To announce these individuals' names probably would be a waste of time. However, it should be mandatory that some permanent record of the individuals jointly acting as unified command be kept for future reference. Finally, there must be a formal announcement of the specific entities represented at the command post.

A word of caution: Initial attempts at unified command could prove awkward and may be downright disastrous. That is why preplanning and disaster drills incorporating unified command are important. Lines of communcation, authority, and responsibility must be established as much in advance as possible. Jealousies and personalities must be left at the door. A disaster scene is not the place or time for avarice, ego, or paybacks. Disaster drills, as much as they are dreaded, are essential if unified command is to be attempted.

ANNOUNCING THE ASSUMPTION OF COMMAND

When formal command is established, it should be announced over the radio along with the initial on-scene report. This radio transmission lets everyone responding (and listening) know from whom to take direction. This fact is vital. Only one individual should be giving direction to incoming units. Too many chiefs spoil the fire! It's true that there is more than one way to fight a fire, but only *one way at a time*, please! Multiple officers with multiple priorities and multiple strategies to get into place can only lead to disaster.

The first officer on the scene shall announce that fact and that he is Command. From that time on, Command is the only officer to make strategic decisions at the fire.[4] More than anything else in the recent history of the fire service, the IMS has made it essential that fire administrators train their officers in how to make strategic decisions at incidents.

The IMS allows chief officers to sit back and watch officers (young and old, new and veteran) take command at all types of incidents. The IMS has given us the opportunity to see who has good strategic concepts and to identify areas of weakness. It also has allowed us to identify and distinguish the team players from the "hot dogs." In short, with the IMS, we have identified some weaknesses that past procedures covered up and we have been directed toward some new areas of needed training.

Table 2-1. COMMAND MODE, UNITS DISPATCHED, ON-SCENE REPORT

Command Mode	Units Dispatched	On-Scene Report Components
Informal	One	1. The fire department is on the scene 2. An initial conditions report (the "picture")
Formal	More than one	1. The fire department is on the scene 2. An initial conditions report 3. The command mode to be used
Unified	More than one	1. The fire department is on the scene 2. An initial conditions report 3. The establishment of unified command 4. Listing of entities represented at the command post

Table 2-2. COMMAND TYPE, UNIT RESPONSE, AND ENTITIES AT COMMAND POST

Command Type	Units Responding	Entities at Command Post
Informal	One	One
Formal	More than one	One
Unified	More than one	More than one

THREE PARTS OF THE IMS

I've heard chief officers from other jurisdictions say, "We don't have time for all that 'junk'; we just go in and put the fire out." I certainly do not agree with that comment. For something that "cleans up" incidents the way IMS does, there's not a lot to it. Basically, the IMS consists primarily of three parts:

- *Command*: Defines the roles, relationships, and responsibilities of personnel responding to an incident.
- *Staging*: The initial response "playbook" of the fire service. It is the area to which responding apparatus report.
- *Sectorization:* Divides the incident area into manageable units or task-oriented assignments.

That's it! You need to define who is doing what—and where. You need a place to put uncommitted units and personnel responding, and you also need a formal way of dividing up the incident. Now, to make this whole thing flow and to enhance the basic premise of the system, we need to address a few additional areas. But there isn't that much to it.

EFFECTS AND BENEFITS OF THE IMS

Change for the sake of change is not only a waste of time, but it displays an air of uncertainty about a department's stability and self-confidence. However, every now and then something comes along that just makes so much sense that you need to take a good look at it. Incident management is that something.

Basically, incident management affects and, if effectively used, improves the following aspects at every emergency situation:

Command

As already noted, the IMS defines the roles and responsibilities of Command and others responding to the incident. By requiring that command be established at the beginning of all incidents, there is no misunderstanding who is in charge of an incident at any particular time. Command must establish strategies and then make the appropriate assignments with the personnel and the equipment already on the scene and en route. Command then can systematically build a specific (preestablished) structure that will effectively meet the needs of that particular incident.

Communications

The IMS establishes a universal form of communication at emergency scenes. Terminology and nomenclatures are preestablished and used at specific times and at certain occurrences. Everyone responding is given an assignment that is specific and known to all on the scene (and those listening). Additionally, Command should control all radio communications. Only certain officers should speak directly to Command, and Command should speak only to certain officers. This follows the chain of command according to the incident's flowchart.

Span of Control/Communications

Incident management sets guidelines relevant to the span of control at emergency scenes. Span of control is defined as "the number of subordinates one supervisor can effectively handle." The emphasis is on *effectively*. At the large 1985 Toledo industrial complex fire discussed in Chapter 1, at one point, 12 companies were venting the roof. At this same fire, more than 24 companies (including recalled personnel) responded—all under the control of one man. That was not the reason the building burned down, but it didn't help. According to the International City Managers Association,[5] the effective span of control at nonemergency fire department operations is 1:12 (officers to subordinates). At emergency operations, the span drops to 1:3-7 subordinates. The generally accepted figure, and the one I will use, is 1:5 subordinates. When using the IMS, this 1:5 figure should not be violated.

A newer concept that mirrors span of control but that has a different twist is "span of communications," defined as "the number of persons (not particularly subordinates) with which one individual can effectively (verbally) communicate." The effective span of communication is approximately the same as that for the span of control. This number differs at an incident according to the place within the incident's command structure and the tasks being performed. Command and the members of Command's staff may effectively communicate with slightly larger numbers of individuals than emergency sector officers. At the command post, Command or Planning, for example, may be able to hold conversations with 10 individuals or more. At the same time, the search sector officer may be able to communicate effectively with only two or three crew members. Span of communication—just as span of control—needs to be addressed at incidents. Unless there is effective communication, the incident will suffer.

Figure 2-4. Command should look at "the picture" in front of him.

THE FOCUS OF COMMAND

Command should focus on the "whole," a theme that resurfaces throughout this book. It should become a part of size-up and continue throughout the incident, regardless of the incident type, size, and complexity. When the first officer pulls up to the incident and assumes command, that officer should look at "the picture" in front of him and then focus on the entire scene. *Command can never afford to get involved in task-oriented actions and should avoid micromanaging specific aspects of any incident.* At a residential structure fire, the focus should be on the entire situation.

At a vehicular accident, the officer should not get involved with patient care. Once Command focuses on specifics, the situation as a whole gets lost. This is a hard lesson to learn. I cite the following example that happened to me when I was "learning" the game.

SCENARIO 1

I was dispatched to a vehicular accident two blocks away. When I arrived, I gave a quick on-scene report:

One vehicle involved with a pole; only one victim at this time. Battalion 1 is Cherry Command. Engine 3, you will be Patient Care.

I walked over to the female victim, who was bleeding severely above the right eye. She was moving her head around, and I was concerned about neck/spinal injuries. I reached in and started C-spine immobilization. At that point, I was finished as Command. (Once C-spine is started, it should not be interrupted until the victim has been boarded.) My ability to "run" the incident was over. It's tough to push the buttons on a portable radio while both hands are on a victim's neck. I lost control of that incident. If a police officer had walked over to me and told me that he had found three children thrown from the car lying in a ditch, what would I have been able to do? I had my hands around this woman's neck.

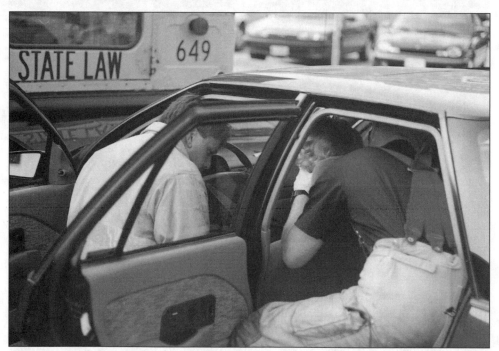

Figure 2-5. A firefighter providing C-spine immobilization to the victim of a vehicular accident. (Photo by author.)

Command's task is to evaluate and prioritize all aspects of the incident and develop a "to-do list"—a list of things that need to be done based on the four priorities of command (explained below)—and make assignments according to that list. Once assignments have been made, a sector officer's responsibility is to accomplish the task assigned using accepted tactics and evolutions. Once this has been done, in some respects, Command's task is complete until the situation changes (determined by Command's own observations or sector officers' reports). This is akin to macromanaging the incident. In effect, the system is based on the following premise: "I'll (Command) tell you what needs to be done. You (sector officers) go do it!" The process should be a continuous process (accomplished by assigning sectors, the subject of Chapter 8).

Once Command allows himself to become focused on a specific activity, other potentially vital aspects of the incident will be neglected. The sector officers are to focus on the specifics. Command must focus on the whole.

ROLES AND RESPONSIBILITIES OF COMMAND

The Phoenix View

One of the early arguments between members of the Phoenix camp and the NFA camp centered around the responsibilities of Command. Chief Brunacini's (Phoenix Fire Department) approach was more simplistic and leaned toward safety. According to Brunacini, "Command (FGC) has the following four responsibilities on the fireground:[6]
- to provide for firefighter safety
- to provide for civilian safety,
- to stop the fire, and
- to conserve property.

The NFA View

The National Fire Academy took a more formal stance and adopted the following responsibilities of Command:
- operations,
- planning,
- logistics, and
- finance.

To me, these responsibilities represent functions more than responsibilities and are components of the organizational structure as opposed to actions such as stopping the fire.

As you can see, the differences are rather striking. Brunacini believed that for his officers to accept the premise of the IMS, he would have to take some of the formality out of the system. It is rather difficult to explain to a fire officer that he must consider finance one of his responsibilities at every incident.

Functions are not situational. An individual's function in life is always present.

Table 2-3. COMPARISON OF RESPONSIBILITES

National Fire Academy ICS	Phoenix ICS
Operations	Firefighter safety
Planning	Civilian safety
Logistics	Stop the fire
Administration	Conserve property

Priorities, on the other hand, are items that should always be considered and reviewed, but how they are applied depends on the situation. As an example, my function as father is to provide a roof over the heads of my family members and to provide for their safety, well-being, and personal and emotional needs. However, my priorities on any given day may differ. Today it's to go to work and then to take my wife out to dinner. Tomorrow, it may be to cut the grass and, on weekends, to watch my son play basketball or my daughter participate in equestrian events.

THE COMPROMISE

Both the Phoenix and NFA sets of responsibilities must be represented in the IMS and must be addressed in some form at every incident, regardless of the incident's type or complexity. In an attempt to keep both lists of responsibilities in the system and to avoid confusion, I offer the following:

The Four Priorities of Command

At every incident, the officer in charge shall, during the size-up process, prioritize actions and assignments utilizing the following four factors:

Provide for Firefighter Safety

Civilians call us to an emergency to solve a problem. It may be a fire, an ill relative, a faulty motor that's giving off sparks, smoke, or a host of other emergencies (real or imagined). To help them solve their problem, we must never take actions that allow us to become part of that problem. If we respond to a fire and one of our own becomes trapped or injured, we will immediately place that "brother" above the situation. That is a fact of life. However, when that happens, we have failed the citizen requesting our help.

SCENARIO 1

I responded to a fire several months ago—a small house with heavy fire in the front living room. Prior to my arrival, an engine officer established command and had the initial line taken in through the front door (tactically, the wrong place to take the line). Crews stepped up onto the front porch, awaiting water.

Before water arrived, a crew member tried to advance just inside the door. At that time, flames blew out of the front door. A firefighter scrambled off the porch with his gear on fire. I watched as crews pulled him to safety, and they began to work on the firefighter. As I stood there trying to pull this scene back together, I noticed the homeowner was standing behind me. He and I watched firefighters cool and undress this firefighter while the home kept burning. I was concerned about the firefighter, but I was embarrassed and angry by our tactics. I never said anything to the owner, but we had messed up!

The first officer on the scene must always ask himself, "Can we enter this structure and operate safely?" If the answer is Yes, then an interior attack can be made or a search commenced. If the answer is No, then we must learn to resist our instinct and change conditions or wait until conditions change themselves. If while at a vehicular accident a member is struck by a passing vehicle or caught in a flash fire while providing C-spine immobilization, then we again will place that "brother" above the situation—and, again, we will have failed the civilian requesting help.

When called to a shooting, stabbing, or the like, we must resist the temptation to rush up, but rather must wait until the police secure the scene. Risks may have to be taken on occasion. There will be fires at which we may have to extend ourselves and do things we normally would not require our members to do. However, those instances should be the exception and not the rule. If we can't operate and be part of the solution, then we become part of the problem and have failed the people who look to us for help.

Provide for Civilian Safety

Once we have determined that fire crews can effectively operate, we next need to provide for the safety of civilians (victims and bystanders). A prima-

ry search should commence when it rises to the top of Command's to-do list. This may be the first item on his list or the last. However, at any structure, a search needs to be done. Its place on the to-do list and the type of search made are situational and are discussed at length in Chapter 12. Bystanders and onlookers need to be kept from harm. That's part of Command's focusing on the whole incident. Command needs to stand back and look at the entire scene, bystanders and all! It is by this focusing on the whole that all aspects of the incident will be addressed.

Figure 2-6. If a firefighter gets injured, we will place that firefighter above the situation.

Stop the Problem

After the safety needs of crews have been addressed and all savable victims have been or are being attended to, our next concern at a structure fire will be to stop the problem. In textbooks I have read on my way up the promotional ladder, the term often used was "confine, control, and extinguish." If the incident is other than a structure fire, say a vehicular accident, our concern might be to extricate trapped victims or pull protective lines for "flash fire watch." If the incident were a service call to check for a smoke odor in a basement, we would find the source and eliminate the problem. "Stop the problem" applies in some form to all incidents except those involving basic EMS (emergency medical services) or ALS (advanced life support), which terminate after the needs of the victims have been addressed.

Conserve Property

Brunacini lists this as the last responsibility of Command. Salvage and overhaul, which will be addressed later, are concerns in this phase. Except for structure fires and a few specific forms of service calls such as frozen/burst water pipes or sprinkler heads or to remove smoke due to a faulty furnace or a careless cook, this phase may not be applicable at all incidents. However, if we respond to a vehicular accident involving victims with minor injuries and no hazardous fluids have spilled from the vehicles, we probably can put a firefighter without dirty turnout gear in the car to do C-spine immobilization and avoid damaging the rear seats.

Likewise, you don't "pop" a door if it can be opened by hand. The first two responsibilities listed above (provide for firefighter safety and provide for civilian safety) should be considered at every incident and be part of the size-up/prioritization process. If we do not do this, then we may become part of the problem and will be of little help to the person who called us for aid.

THE FUNCTIONS OF COMMAND

As a line battalion chief, when I'm not on an emergency incident, I have specific things or functions for which I am responsible. I'm responsible for the firefighters on "A" platoon. It is my task to set up and oversee training for the members in my battalion. In simpler terms, I view my function at nonemergency

Figure 2-7. At this incident, "stop the problem" equates with putting out the fire.

operations as "making sure that the officers assigned to my battalion are doing their jobs." When an incident occurs, I must ensure that they can participate effectively in the *operations* section of the department. I must also *plan* for my crew's training, building inspection, preplanning, and so on. I must make sure that their *logistical* needs are met and that they have the appropriate tools and equipment to do their jobs. Finally, I must act as a liaison between the *adminstration* and the line. (By the way, I believe the latter to be the most difficult part of being a line battalion chief. You have to learn to walk the line between being a line firefighter and part of the administration. Both roles are necessary, and a balance must be achieved if you are to be successful. You need to channel the adminstrative aspects of the job by being the "man in the middle.")

At emergency operations, Command shall be responsible for the following four functions at every incident:

- *Operations: Mitigating the emergency.* At a fire, he is responsible for ensuring that the fire is extinguished. At a vehicular accident, he is responsible for patient care, extrication, and so on.
- *Planning: Assessing previous, current, and future needs of the incident.* At a fire, he should consider construction type, occupancy, fire travel, and weather conditions (to name a few) and plan accordingly. At a hazardous-materials incident, he must consider the materials involved, topography, weather, and risk (among other factors) and then plan accordingly.
- *Logistics: Ensuring that the necessary personnel, equipment, and tools are at the scene.*
- *Administration: Ensuring that specific administrative aspects of the incident are attended to, including cost analysis, overtime, and even legal considerations.* If the incident begins to expand, Command can delegate these functions to other officers (normally chief officers) already at the scene or responding.

Each of these functions will be discussed further in Chapter 6.

The individual in charge must do other things at a structure fire or other incident as well. Determining the owner and occupant and the cause of the fire come to mind. Merely handling the above-mentioned functions will not necessarily "meet the needs" of the incident. Procedure should dictate the specific items that need to be handled by the officer in charge of incidents.

WHAT INCIDENT MANAGEMENT DOES

As already stated, the IMS allows one individual to have total control over an incident. It eliminates multiple or revolving commanders and calls for strong leadership from the onset of an incident. It makes "thinkers" of the first-arriving officers at the scene of an emergency and identifies poor strategists. Prior to the IMS, as long as a company was doing "something" when the chief arrived, things were presumed to be going well. With the IMS, the first-arriving officer must set strategies that may carry through the entire incident. Assignments are made from the onset, and everyone responding knows what is already being done and by whom. Once the initial officer is relieved by a higher-ranking officer, those strategies are discussed and normally built on. This requires that all officers have a good knowledge of strategic concepts and the capabilities of the responding companies. In short, the IMS develops better first-level officers.

Incident management provides for expendability at any incident. First, the system offers a simple method of determining who is in command and for making a smooth transition of command to higher-ranking oficers if the situation dictates. (This will be explained later in Chapter 3.) Additionally, incident management makes it possible to use more than 200 companies without violating the recommended span of control at an incident.

Incident management allows for input from experts in specialized fields, such as haz-mat, or the use of HVAC (heating, ventilation, and air-conditioning systems). While the possibility existed for input from experts prior to the inception of the IMS, the system has a specific place for it in a controlled atmosphere. This should stop a chief officer from asking questions relevant to technical areas unrelated to stategies and tactics for egotistical reasons. No longer will a chief have to pretend that he has a grasp of HVAC or the effects of halon on semiconductors!

Incident management is designed to let you (as Command) pass off specific obligations or duties to others while you (as Command) still maintain control. Delegation is a large part of the IMS. You can't do it all. The IMS allows you to assemble your "team" at the command post and draw from its expertise, which affords you time to maintain "focus" on the big picture.

Figure 2-8. The incident management system allows for input from experts in specialized fields.

WHAT INCIDENT MANAGEMENT IS NOT

- *The IMS is not designed to take authority away from the chief having jurisdiction.* In fact, it gives him more control over the incident. Primarily this statement concerns itself with smaller departments and mutual-aid responses. I do considerable teaching for small departments in the area. I constantly hear of the reluctance of these chiefs to initiate and use the IMS. They believe that the system calls for an onslaught of nosy borderland chiefs to swarm around the command post like flies on last week's garbage. The IMS defines responsibilities. If assignments are given to these chiefs and expectations are defined, then little time can be left for "swarming."

- *The IMS is not designed to set strategies or control tactical evolutions.* Nothing in the IMS says how fires should be extinguished. Several months after the implementation of the IMS in Toledo, and following a rash of large vacant warehouse fires, a few officers were overheard saying, "We wouldn't have lost those buildings if we hadn't been using the incident command (management) system." Hogwash! Buildings are lost to fire for only two reasons: The chief and company-level officers used poor strategic or tactical operations, or the fires simply overwhelmed resources. All the strategies and tactics would be insignificant if the water needed to absorb the Btus was not there when the structure was still resisting gravity. (You, for example, may encounter a fire in a structure that contains hazardous materials and may lose the structure because of the materials involved, their reaction with extinguishing agents, or your inability to get the appropriate extinguishing agents on-scene fast enough. I consider this a case of the "fire's overwhelming you.")

  Note that in both of these reasons, no mention was made of organization or communication—two major parts of the IMS. The "system" does not tell companies to "lay in," vent, commence salvage, extricate victims, or set up defensive operations. Command (the person) is responsible for that. Remember, the IMS is a management tool. It defines responsibilities and nomenclature. If the system is in place and Command issues no orders, then nothing will be done, and the structure will burn to the ground. That's the way the system is designed. No one does anything until Command gives the order. Otherwise, control would be lost and responsibility cannot (and should not) be placed on the officer "in charge" because if there is freelancing, he really is not "in charge."

- *The IMS is not complicated.* To put it as simply as possible: "One person is in charge of every incident, and no one does anything until he or she says to." That's it. Putting the above statement into effect at a third-alarm high-rise fire in downtown Toledo, however, is not as simple as it may sound here. But the basics of the system are not complicated.

Figure 2-9. A chief officer assigned as "Sector 3" communicates to crews working in his sector.

FLOWCHARTS

Flowcharts are important for depicting the roles and relationships of fire-fighters responding to incidents. They allow Command to visualize the span of control for the incident and aid in making adjustments to overtaxed areas of the incident. Command should diagram any incident larger than a regular alarm response. Flowcharts are presented in other sections of this text, including the last chapter.

Another form of command flowchart is below. In this chart for a normal structure (house) fire, Command has specific sectors under him. Note: Command is still responsible for the four functions at every incident. If

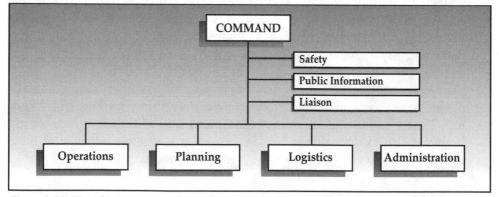

Figure 2-10. This "basic" command flowchart indicates the lines of authority between Command and Command's functions and staff.

Command does not assign an officer to those functions, then Command is responsible for them. Boxes in dotted lines are the "things" for which Command is still responsible. I call these "ghost" sections. Boxes in solid lines are sectors that Command has assigned to others. This flowchart is an acceptable chart for a house fire.

Either flowchart is acceptable. The first depcits the formal IMS. Sector assignments would be under Operations. The second depicts a newer chart that uses both. More will be said about this in later chapters.

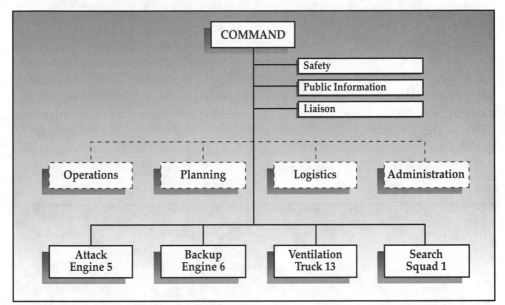

Figure 2-11. Flowchart indicating functions of Command and several sector assignments.

COMMAND PRESENCE

In my lectures, the discussion of command presence often gets the most attention. Most officers can put themselves in the positions I discuss and remember past experiences where the chief in charge totally lost it and, in turn, the situation was lost.

More than anything else, command presence is how Command appears to others. If you are an officer who can be put in the position of taking command of an incident, you must realize two things: You will set the tone of the incident, and you are being watched. How you handle the initial stress of the incident in great part will determine how the incident will eventually turn out.

My father was an excellent golfer. My grandfather was a professional golfer. I had a cutoff set of wooden clubs when I was five years old. As in the fire service, the arena of golf abounds with tradition. I was taught at an early age never to wear shorts or cutoffs (that was a '60s thing) on the golf course. "You never see pros on TV golfing in shorts or not wearing a shirt, do you?" my dad would ask. He would always end that statement with, "You've got to look good to play good!" That certainly holds true on the fireground as well. However, on the fireground, the "look" in the above statement relates less to clothes and more to actions.

Thoughts come to mind of one of my old deputy chiefs throwing a pike pole (why he had a pike pole in his hand we never knew) into the ground like a spear when a firefighter fell through a hole (that everyone knew was there) on the second floor of a vacant house, or of a chief officer running around at a fire, going up ladders to check on the size of vent holes or sticking his head inside the smoke coming from an open door. It's hard to instill confidence in the crews operating at an incident when the one in charge can't control his own actions.

You have to learn to—no, force yourself to—remain calm at every incident to which you respond. If you are excited, then your crews will tend to get excited. If you get flustered, they (and the incident) get flustered. Step back (actually) and look at the whole picture. Focus on the large problems, not individual traumas. Make clear, focused assignments. The more confusing the incident is (and the later in the incident you arrive), the more slowly and directly (mild volume) you need to speak. You need to get your officers' attention without yelling. Above all, set the tone: strong, direct, focused. Command!

Figure 2-12. Command presence requires that Command remain calm and focused at every incident.

Endnotes

1. Brunacini, Alan V. *Fire Command*. 1985. National Fire Protection Association (NFPA) Publications, 4.

2. *Standard on Fire Department Occupational Safety and Health Program*, National Fire Protection Association, 1992.

3. Sit-Stat officers maintain an up-to-date reporting of the incident and its status. Part of this officer's duty is to keep an acccurate accounting of what is taking place at the incident.

4. This does not hold at larger incidents when the position of operations officer is established. Chapter 5 covers the responsibilities of operations extensively.

5. *Managing Fire Services*, Second Edition. 1988. International City Managers Association, Washington, D.C.

6. Brunacini, 4.

REVIEW QUESTIONS

1. The incident management system (IMS) is a _____ method of operating at _____ incidents to which the fire department responds.

2. The IMS defines the _____ and _____ of all units that respond to an incident.

3. The IMS stipulates that one individual control an incident, thereby eliminating all forms of _____.

4. What are the components of the basic premise behind the IMS?

5. Define "Command."

6. What the individual in command of the incident is called does not matter. What matters are the following:
   (a) _____
   (b) _____
   (c) _____

7. What Federal Act mandates that the IMS be used at all haz-mat incidents?

8. The type of command used depends on what two factors?

9. What are the three types of command?

10. For each type of command, list the following:
    (a) the number of officers responding.
    (b) the number of entities represented at the command post.

11. What are the three components of the IMS?

12. The IMS affects and thus improves which three aspects of every incident?

13. What is Command's task at every incident?

14. According to the author, what are the four priorities of Command?

15. According to the author, what are the four functions of Command?

DISCUSSION QUESTIONS

1. Define "freelancing." Review your department's operations to determine whether freelancing is a problem in your department.

2. Discuss the concept: "Command shall be ultimately responsible for the outcome of the incident" as it relates to interior operations at a fire. Discuss "constructive knowledge" and "reasonable, prudent" actions.

3. Reflect on recent incidents to which you or your department have responded and review them from the perspective of effective span of control.

3

Establishing Command

As we have seen, the incident management system (IMS) demands that someone be in charge (Command) at every incident, all the time. Let's move from the "what" of command—its definition and the roles of command—and focus on the "when" and "how" of establishing command.

KEY CONCEPTS

✓ Command mode
✓ Conditions report
✓ Establishing command
✓ Fast-attack mode
✓ Nothing-showing mode
✓ On-screen report
✓ Passing command

WHEN TO ESTABLISH COMMAND

Not all organizations agree on when command should be established under the IMS. Some say command should be established only when a second alarm is sounded. Others believe that the first chief officer on the scene should take command on arrival. Still others believe the first officer on the scene should take command until relieved by a higher authority or until the incident has been terminated. In some departments, standard operating procedures (SOPs) deter-mine the actions of the first-arriving units, but I find it hard to believe that all potential scenarios are covered in these SOPs.

There should be no happy medium here. Command should be established as soon as the first officer (or unit) arrives on the scene. In this way, a strong command presence is established from the onset of the incident.

The concept that "someone shall be in charge of every incident to which we respond" is vital to effective, efficient emergency operations. It is never acceptable to have to make an assumption about who is in charge of the incident when more than one officer is present. Multiple commanders can lead to multiple objectives, strategies, and proposed outcomes. Confusion likely will be the outcome. Management of an emergency scene is most effective when a strong command presence is in effect from the beginning. This command presence can be strengthened as the incident expands, but all at the scene and en route should know from whom to take direction.

NFA Perspective

Fire organizations that followed the National Fire Academy's form of the IMS say the first-arriving officer (or unit) should establish command (be Command). This individual then should establish a command post and run the incident from there. This approach ensures a strong command presence from the outset of the incident.

Phoenix Perspective

The Phoenix form of incident management advocates a more flexible, situational approach to establishing command. Proponents of the Phoenix system believe it unrealistic to require the officer first arriving at a house fire and finding a mother out in front screaming "My baby, my baby!" to stay out in front of the house directing incoming companies with a portable radio in his hand.

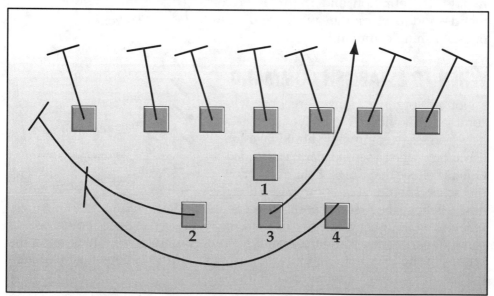

Figure 3-1. 32 dive right.

COMMAND MODES

To address such a situation, Chief Brunacini developed *command modes* that give the first-arriving officer options with regard to initial actions at an incident, depending on conditions.[1] The Consortium, established to resolve the differences between the Phoenix and NFA forms of command, also recognizes these command modes, which it refers to as *command options*.

Command modes are selected according to the conditions of the situation. I refer to command modes as the playbook of the fire service. When the quarterback (Command) calls the play (command mode) according to what he sees (the situation), then the team (other units responding) knows what to do.

When I played high school football (I am going to date myself with this) and the quarterback called a "32 dive right," the third man in the backfield (the fullback in a "T" formation) would take the handoff and run through the "#2" hole (between the guard and the tackle) on the right side. Everyone else would block accordingly. The quarterback did not have to tell every player what to do on every down.

Command modes are similar to that. When the first officer arrives and reports "nothing showing," all responding units now know what to do. Command modes should be part of Command's on-scene announcement.

Remember: Command modes are situational—they are dependent on what the officer first observes on arrival. Also, command modes can be used for most, if not all, types of incidents—not just for fires.

Types of Command Modes

The command modes are *nothing showing*, *fast attack*, and *command*.

Nothing Showing

More than 90 percent of all fires are handled by one line or less. At the majority of the fires to which we respond, the first unit on the scene sees no evidence of a fire on arrival.

Figure 3-2. If the "picture" looks like this on arrival at the scene of a reported structure fire, Command would be expected to use the nothing-showing mode.

The first-arriving unit gives an on-scene announcement including "Nothing showing." The officer and his crew investigate. At that point, all other responding units stage (stand by) at an appropriate location.

In this mode, the first-arriving officer observes no smoke or other signs of fire on arrival. The officer should state this fact along with a repeat of the address and unit designation.

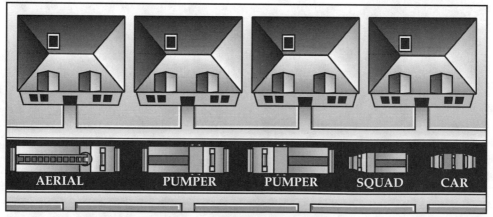

> *Engine 5 is at 1945 Vermont. Nothing showing.*

At this time, all units responding or listening to the transmission know two key things:
- A fire department unit has arrived at the scene.
- There are no apparent signs of a fire at this time.

In the nothing-showing mode, we would expect the officer to participate with his crew in determining the nature of the incident. They would leave the apparatus and investigate to determine if the fire department's services are needed. It is assumed that the first officer on the scene is in command and is going to investigate. The fact that this first unit is Command and that it is going up to investigate need not be announced over the radio. That is what is expected.

Here comes the playbook.[2] Once all responding units hear the first unit use the nothing-showing mode, they stage until directed by Command. No other units shall go up to the incident without being directed to do so by Command (the first officer on the scene).

I'm not sure how that sits with the firefighters in your jurisdiction, but in Toledo it was a monumental task to keep rigs from going up to the scene in the nothing-showing mode. Firefighters seem to have this desire to help. On the face of it, that's not a bad trait, but there is really no sense in having three engine companies, a truck, a heavy squad, and a chief officer pull up in front of a bungalow on a narrow street when there probably is no fire.

Figure 3-3. Engines and trucks positioned on a narrow street.

Up until the Toledo Fire Department instituted incident command, we all would drive right up to the structure in question. I've seen shoving matches among drivers and officers over who was going to back up so that the other rigs could pull ahead after a false alarm or food-on-the-stove call. This type of action falls under the command-presence category previously discussed. If the first unit goes up to a house and the other units stage at intersections, then we "look important." Not excited? However, what picture do we paint in the minds of civilians (and the children who love us) if we all pull in front of the structure and then argue amongst ourselves over backing up?

Fast-Attack Mode

The first unit arriving at the address gives an on-scene announcement including the term "fast attack." The officer and his crew participate in the needed activitites. The next unit on the scene establishes formal command. Once formal command is established, all other responding units stage.

Here's how it works. The first officer on-scene announces that he is on the scene, gives a condition report, and advises that the fast-attack mode is in use.

Engine 5 is at 1945 Vermont. Light smoke is showing.
Engine 5 is going fast attack.

Figure 3-4. The first unit on the scene of this structure fire used the fast-attack mode. The officer believed his participation in stretching and advancing lines made a difference in the incident's outcome.

At this time, all units responding or listening to the transmission know four key things:

• A fire department unit has arrived at the scene.

- There is some type of emergency.
- The first officer is joining his crew to participate in the incident.
- The next-in unit on the scene (regardless of its type) establishes formal command. Once formal command is established, all other responding units stage.

In the fast-attack mode, the officer feels that his two hands will have a definite impact on the incident's outcome—in other words, if Command participates with this crew, the incident probably will be handled and probably no additional units will be needed. The converse is also true. If the officer does not participate and opts to take formal command, it is likely that the incident will escalate into a working incident. For this mode to be effective, the first officer must consider several variables in a relatively short period of time, including the following:

The extent of the incident on arrival. Will the officer's participation have a definite impact on the incident's outcome? With a small fire in the kitchen or in a wastepaper basket in the bathroom, this probably will be the case. However, if there is a totally involved vacant house in the middle of a field, then the officer would be hard-pressed to explain how his two hands will have an impact on the outcome. The house is coming down whether or not he participates.

The location of other units responding. If the unit operates in the inner city in an urban area and other units responding are just seconds behind the first unit, then the time lag prior to the next incoming unit's establishing formal command will be short. Additionally, if definite rescues are to be made and additional responding units are some distance away, the officer may use the fast-attack mode to participate in the rescue. (There are no clear-cut answers here. Procedures and recommendations from your department chief should guide you as to when to use fast attack.)

The next officer to arrive on the scene goes up to the incident and establishes formal command.

> *Engine 23 is at 1945 Vermont. Engine 23 is Vermont Command.*

At this time, all other responding units stage, unless the responding chief has not yet arrived on the scene. Normally, chief officers should take formal command of any working fire or incident. When the chief arrives, he should go up to the scene and take command from the officer.

Note: The officer going fast attack is still responsible for the incident until the next-in officer establishes formal command. If the officer gets to the top of the second-floor stairs and finds that the stairway can't be made because of the heat, the officer should direct the truck (or someone) to vent the roof.

The fast-attack mode will not last very long. It will end in one of three ways:

- The first-in officer (the one going fast attack) goes in and sees that the incident is small in scope and announces that he can handle the incident.

Engine 5 can handle 1945 Vermont. All other units can go back in service when ready.

That announcement puts the command structure back down to informal command with only one officer at the scene.

- The first-in officer (the one going fast attack) gets inside and determines that the fire is beyond the fast-attack stage and decides to go back outside and establish formal command.

Dispatch, this fire has gotten into the second floor. Engine 5 will be Vermont Command.

At this point, the officer has established formal command; all incoming units should stage, awaiting direction.

- The next-in officer establishes formal command.

Dispatch, Engine 23 is at 1945 Vermont. Engine 23 will be Vermont Command.

At this point, the officer has established formal command; all incoming units should stage, awaiting direction.

Command Mode

The first unit reports to the address. The unit officer announces he is Command. Once someone formally establishes command, all other responding units stage and await direction from Command.

In the command mode, there is a working incident. The first officer has determined that it would be more beneficial to the incident to stay outside and direct other incoming units than to particpate in the attack.

Engine 5 is at 1945 Vermont. We have a house totally involved, with exposure problems. Engine 5 is Vermont Command.

At this time, all units responding or listening to the transmission know three key things:

- A fire department unit has arrived on the scene.
- This is a working fire.
- On their arrival at the incident, they should stage in an appropriate location until directed by Command.

Figure 3-5. An example of a situation that required the command mode. What difference would the officer's two hands make in the outcome of this incident?

Table 3-1. COMMAND MODE, UNITS ON-SCENE, ANNOUNCEMENT

Command Mode	Number of Units Dispatched	Items Included in On-Scene Report
Informal	Less than one	1. Fact that fire department unit is on-scene
		2. Initial conditions report
Formal	More than one	1. Fact that fire department unit is on-scene
		2. Initial conditions report
		3. Required command mode
Unified	More than one	1. Fact that fire department unit is on-scene*
		2. Initial conditions report
		3. Required command mode
		4. Listing of the entities represented

*More than likely, the incident progressed from informal to formal to unified or from formal to unified command. Hence, fire units probably already will be on the scene, and this item may not be required.

PASSING COMMAND

In formal or unified command, the individual in charge can be changed if the need arises. This process, referred to as *passing command*, applies to changes up or down the chain of command. Normally, a logical progression up the promotion ladder is followed. As the incident starts to de-escalate, this tends to reverse itself; command is passed down to members with lesser rank. Sometimes, however, passing down command cannot and should not be done. As in the case of command modes, when to pass command depends on the situation.

When the first unit arrives at these incidents, the officer should issue a radio report identifying the unit, briefly describing conditions, and noting the proper command mode to be used.

Engine 7 is at 2315 Cherry. We have light smoke showing. Engine 7 is Cherry Command.

The other unit(s) responding now know the following:
- Engine 7 is on the scene.
- There is some sort of fire (Engine 7 reported "light smoke showing").
- The officer of Engine 7 has taken formal command of the incident.

Only Command (Engine 7 officer) can make additional requests for equipment, terminate the incident, or assign companies. The situation will remain that way until the incident is over or someone formally takes command from Engine 7 officer.

When to Pass Command

When to pass command has always been a question. If a chief officer is dispatched and not canceled, should that officer automatically take command on arrival?

Brunacini provides a clear answer that may be followed as a basic rule of thumb: *"If you can't improve the quality of command, don't transfer it!"* If the fire is small and the officer in command is acting properly, there is no reason to pass command. Egos should not play into this issue. I realize this statement is open to interpretation and is not based on objective criteria.

SCENARIO 1

An engine with a newly promoted lieutenant and an engine with a 20-year captain riding the seat are dispatched to a garage fire. The engine with the young lieutenant arrives first and finds a garage totally involved and the fire spreading to a nearby house. The lieutenant fails to lay a line in to protect the exposed house. If the captain responding observes this tactical oversight, is the captain obliged to make the necessary "adjustments"?

This situation might be handled in several ways:

- The captain could report to the scene and formally take command.
- The captain could report to the scene, approach the lieutenant, and make some specific suggestions. (If this scenario were to occur, more likely than not, the reason for the tactical oversight could be nervousness, and a little "coaching" from the captain probably would take care of the problem.)
- The captain could prompt the lieutenant by radio with statements such as, "Engine 3 to Command: Do you want us to lay in and take the exposure?" This, again, would probably take care of the problem.

After the garage fire, the captain could talk to the lieutenant about the situation and hopefully soothe the lieutenant's nerves. When they get back to the station, the captain may want to discuss tactics with the lieutenant.

As stated above, personalities should not play into the passing of command. The fewer times the "ball is passed," the less the chance of a fumble. Pass command only until the situation and its specific needs are being addressed.

An Alternative Method

I thought about this for hours when developing procedures for my department. Another way to determine whether command should be passed and one that removes the "ego thing" is by asking, *Are the needs of the incident being met*? This is a subjective statement. However, it is a much softer statement than "If command can be improved." What better way to determine when command should be passed than by looking at the scene and determining if "its" needs are being met? Such needs would include the following:

- Are all the appropriate sectors assigned and in good working condition?
- Is the incident still expanding or beginning to de-escalate?
- Are the needs of the occupants being addressed?
- Are the needs of fire crews (such as rehab) being met?

If all of the above items are being addressed, then I feel there is no need to step in and take command from an officer of lower rank.

If the statement "If the needs of the incident are being met, then command need not be passed" is correct, then so is the statement "If the needs of the incident are not being met, then command should be passed." This certainly will be a judgment call.

As mentioned in the previous chapter, the IMS is expandable. As the incident grows, so can and should the command structure. The idea is to address the needs of the incident, not play musical chairs. Passing for the sake of passing (or ego) is burdensome to the incident and serves no purpose. If the needs of the incident are being met, then command need not be passed.

Note: One final factor must be considered, however. Our staffing is being attacked on a daily basis. "Do more with less" is the common cry from city hall. To effectively and safely handle the incident, it may be more prudent for the responding chief to take command at a working incident and send the initial commanding officer back to his crew to "beef up" the attack company even though the needs of the incident are being met.

SCENARIO 2

A three-person engine company arrives at the scene of a single-story home with a small bedroom fire. The officer establishes command and assigns the remainder of his company and the next-in company to attack. The chief arrives, observes the conditions, and takes command from the officer. The chief assigns a truck to ventilation and releases the remainder of the assignment.

The initial command officer probably could handle the incident. However, no one will dispute that a three-person engine company is not too efficient. The initial commander was correct in assigning the remainder of his company to join in with the next-in company, thereby "beefing up" the attack company and still meeting the incident's needs. The chief was the next officer to arrive and was equally correct in taking command from the initial commander and allowing the officer of the first-in unit to join up with his crew, thereby eliminating the need for the second-in engine.

When command is going to be passed, this fact should be announced over the radio—again, to let all units dispatched or responding and those listening know from whom to take direction. Additionally, it indicates who the "responsible" party is. If errors in judgment are made, it will be easier to identify who made the error. However, the IMS is not a vehicle for identifying error and placing blame. It simply defines responsibility.

Dispatch: Battalion 1 will be Cherry Street Command.

Now the responsibility of the incident has been formally shifted to Battalion 1.

Someone has been in charge of the incident since the first unit arrived.
Note: The name (Command) never changes. The people taking command may change, but "Command" is always the one in charge.

If the incident escalates to a multiple alarm and the deputy chief, division chief, or chief of department arrives, neither can nor should give any direction at the fire until that officer formally takes command. Now, we all know that officers will come to the command post and give Command all the "face-to-face" direction they deem necessary. But Command should be the only one giving crews operating at the scene direction over the radio. The one exception to this is explained in Chapter 6.

When the incident starts to de-escalate and command is passed back down to the company-officer level, a formal announcement should also be made.

Dispatch: Battalion 1 will be coming in service leaving the fire. Engine 7 officer will be Cherry Street Command.

Again, there has been a smooth transition of command while maintaining one person in charge.

Passing Command to an Outside Agency

At times, command should be passed to an outside agency. At a vehicular accident, for example, I encourage my officers to pass command to the police, who normally are still on the scene taking their reports after the victims have been transported and fire crews have picked up. I believe this is necessary from a legal perspective. Passing command makes it possible to document when we left the scene and the agency we left in control. In this way, any actions taken by city or outside agencies cannot be attributed to the fire department.

I tell my officers to walk up to and notify the police officer (or representative of any other agency involved) that we are leaving the scene and that that agency now has control or command (or any other term that puts the agency on notice that it is now responsible for the scene). I also suggest that the officer transmit that fact:

> *Dispatch: Engine 5 is leaving the scene on Main Street.* TPD (Toledo Police Department) *is now Main Street Command!*
> *Dispatch* (giving a time check): *Okay, Engine 5, at 1343 hours.*

This procedure also may be used when leaving the scene of a hazardous-materials incident when the scene is turned over to an agency such as the Environmental Protection Agency or an outside cleanup contractor. I believe we "legally" need to put the remaining entity on the scene on notice that we are leaving the scene and it now is responsible for whatever happens. Discuss this point with your Legal Department for a final determination.

INCIDENT TIME LINE

The sequence in which individuals have taken command and the actions that have occurred during the incident can be plotted on an incident time line, basically horizontal graphs (Figure 3-6). If an incident involves an arson conviction or civil litigation and goes to trial, the attorney often will make an incident time line to illustrate specific occurrences at the incident. Incident time lines are useful for visualizing what took place and when at an incident.

USE OF CREW

I am often asked the following two questions when teaching the IMS:

Q. *How can Engine 7 take command and then assign Engine 7 attack?*

A. We have determined that to have a strong command presence from the onset of the incident, the first officer on the scene should establish com-

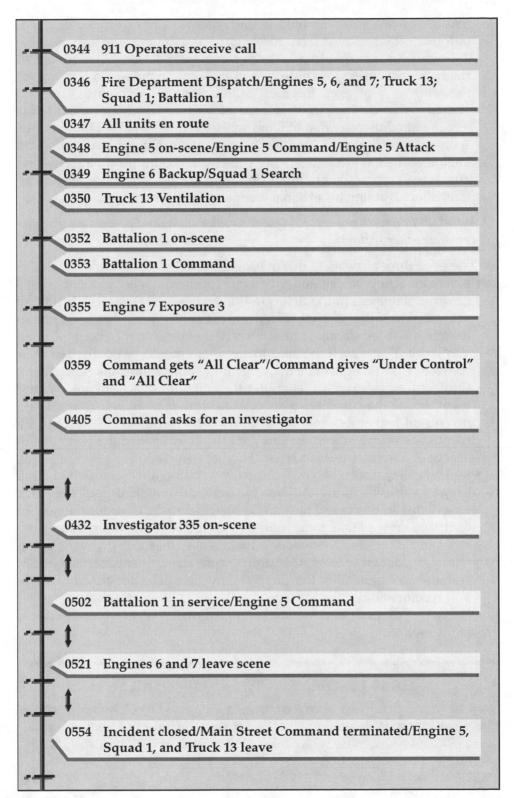

0344   911 Operators receive call

0346   Fire Department Dispatch/Engines 5, 6, and 7; Truck 13;
        Squad 1; Battalion 1

0347   All units en route

0348   Engine 5 on-scene/Engine 5 Command/Engine 5 Attack

0349   Engine 6 Backup/Squad 1 Search

0350   Truck 13 Ventilation

0352   Battalion 1 on-scene

0353   Battalion 1 Command

0355   Engine 7 Exposure 3

0359   Command gets "All Clear"/Command gives "Under Control"
        and "All Clear"

0405   Command asks for an investigator

0432   Investigator 335 on-scene

0502   Battalion 1 in service/Engine 5 Command

0521   Engines 6 and 7 leave scene

0554   Incident closed/Main Street Command terminated/Engine 5,
        Squad 1, and Truck 13 leave

Figure 3-6. Time line.

mand. However, the minute the officer says, "I am Command," he is automatically removed from his original assignment—let's say Engine 7. Command then should establish a command post and begin to size up the incident, develop a strategy, and make assignments to accomplish the strategy. However, Engine 7's crew and the apparatus can and should be used. The size and type of unit to which Command was assigned and the situation at hand will determine what should be done with the remainder of Engine 7's crew. If Command was riding an engine and enough members are present to mount a safe attack—and attack is Command's first item on the to-do list—then Command can (and should) assign his original company to attack.

Q. What do I do with my crew if I have to take command or become the officer of another sector? [3]

A. There are three answers to this question.

- Name an acting officer, and give your company an assignment. That's how the statement in the first question comes to make sense. Who will act as officer if the original officer must take command has been predetermined. We require that our officers designate an actor every morning at roll call. That way, if the need arises, there should be no discussion or confusion as to who will be in charge if the officer must take command. Contractually, we pay an actor after the first hour of acting time. We use members on the current promotional list (previous members if there is no current list) or delegate by seniority if no members are on a list. By naming an acting officer, we can send the crew inside with someone in charge of and responsible for the action of that crew.
- Link the remainder of your company with another dispatched unit. Consider this if staffing on your apparatus is low or if personnel lack experience. In this case, lines can be stretched to the door for the next-in company crew to take in as soon as they arrive at the door. Inform the incoming officer and receive confirmation over the radio if you are going to assign your crew to another unit. If your department uses a check-in system of accountability, this practice may have to be fine-tuned to maintain accountability.
- Use your company as command staff. Following is my classic example of this option.

USING COMMAND'S CREW AS COMMAND STAFF

Your unit is first to arrive at a confirmed high-rise fire. One unit member begins writing down assignments (for the command board when a chief arrives), and another firefighter takes a clipboard and establishes a "lobby sector" until another officer arrives.

The driver could proceed to the sprinkler or standpipe connection and start to hook up.

These are suggestions. The assignments can be varied, or a combination of the above can be used. Situations and ingenuity are the keys.

WHAT TO DO WITH THE CHIEF?

An age-old question is, What do we do with the chief? Think of the ego problem that could be associated with the following statement: "If command cannot be improved, then command need not be passed." Who determines whether command can be improved? Should we interpret the responding chief's getting canceled as an indictment of the chief's ability to run an incident?

First of all, R.H.I.P. (rank hath its privileges). If the chief wants to come to a rubbish fire and take command of a single company attacking a dumpster, that is the chief's call. (It certainly says something about what he feels his time is worth!) We have a saying in Toledo: He who has command does the paperwork—NFIRS (National Fire Incident Reporting System)[4] and all! Now, if a chief happens to get dispatched to a rubbish fire (911 operators can make mistakes) and wants to hang around and watch a company put out the fire, then there's not a lot we can do about it.

Our procedures require that the dispatched chief take command of all working fires to which he is dispatched. He may opt to have the original Command keep command as a training session if the member is in a promotable position, as mentioned earlier. If so, this fact should be announced and noted over the radio.

In our organization, only battalion chiefs work the line. Deputy chiefs fill staff positions. They are dispatched only to second-alarm or larger incidents. When they arrive at a large incident, our procedure states that they will take command of the incident. The original battalion chief is moved to operations chief. This is explained in Chapter 6.

Endnotes

1. Brunacini, Alan V. *Fire Command.* 1985. National Fire Protection Association (NFPA) Publications, 30.

2. If an engine company is not the first unit on the scene, the Toledo Fire Department procedurally allows an engine company to proceed up to the incident in the nothing-showing mode in case there is a small fire and water is needed from the tank. All other responding units stage in the nothing-showing mode.

3. Brunacini, 32.

4. Fire reports with specific information and nationally used codes that provide statistics related to the national fire problem.

REVIEW QUESTIONS

1. What three items should the first officer on the scene include in his on-scene report when operating in formal command?

2. Complete this statment: Dispatchers should take direction only from
_____.

3. To what needs are we referring when we say "command should be passed on the basis of whether the needs at the scene are being met"?

4. It is accepable for a higher-ranking officer to give direction to units on the scene as long as he gives his rank first during the transmission.
___ True ___ False

5. When the incident starts to de-escalate, command can be passed back down the chain of command to the company officer level.
___ True ___ False

6. What determines which command mode should be used at an incident?

7. Command modes should be part of which report?

8. Why are command modes situational?

9. In the nothing-showing mode, what should the officer and his crew do?

10. Once the officer reports "Nothing showing," what should all other responding units do?

11. In the fast-attack mode, who should establish formal command?

12. The fast-attack mode will last only a short time. In what three ways might it end?

13. Before using the fast-attack mode, what two things must the officer consider?

14. What are the three command modes?

15. In what three ways can a command or sector officer's crew be used?

DISCUSSION QUESTIONS

1. Why is it important that the assumption of command be announced over the radio?

2. Share with fellow students/department members past experiences of multiple commands at incidents and the associated problems encountered.

3. The text discusses a garage fire at which the second-due officer saw tactical problems with the initial command. Can you see that happening in your department? What do you believe would actually happen if this situation arose in your department?

4

The Command Post

TOPICS

- **Command post**
- **Mobile command**
- **Types of command post**
- **Relationship between the type of command post and the type of command**
- **Suitable areas for a command post**
- **Drive-bys**
- **Who should operate at the command post**

The command post is the area at which Command is expected to operate. Usually, Command will operate at or near his vehicle or, in the case of a single-family structure fire, at or near the first-in unit (engine, truck, or squad).

The incident commander (IC) determines the location of the command post. It is extremely important that Command remain at the command post for the majority of the incident so that he will be easy to locate. Before the incident management system (IMS) was implemented, chief officers (or others in command) would circle the fire area.

KEY CONCEPTS

✔ Command post

✔ Drive-by

✔ Duration incident

✔ Formal command post

✔ Mobile command

✔ Mobile command post

✔ Stationary command post

Some chiefs say that that was "to get a good, continual look at the structure." I believe that, for the most part, circling was a release for nervous energy. I can recall countless fires at which we counted the number of times a chief (or chiefs) passed us as we dumped water on defensive fires. I can also recall countless times when I looked for the chief to give him a message I did not want to give over the radio and could not find him. Usually, he was on one side of the building or on the second floor while I was looking in the area he was "last seen passing by." This was an unnecessary waste of time and energy for all concerned.

I have read fire textbooks and articles that say the most effective place for a chief officer is inside the fire building, right behind the nozzle. I find it hard to believe that a chief officer can effectively focus on all the needs of an incident when inside a structure. Command must trust his officers! Command must stay outside and maintain focus on the whole incident while the officers do their jobs.

The location of the command post generally depends on the type of command established. Command determines its location. However, some guidelines can help ensure that you select the best type of command post and location.

GOING MOBILE

I have been a chief officer for more than 10 years and have directed incidents with and without incident management. I have circled buildings in my day and sometimes still fight the temptation to circle buildings. In Toledo, we borrowed a term used in Phoenix: "Command is mobile,"[1] which means that Command is leaving the command post for a short time. The fact that Command is going mobile should be announced over the radio so that companies on the scene and listening know that Command will not be at the command post.

Figure 4-1. A chief officer going mobile.

 Noble Street Command is mobile. Going to 1344 Noble to talk to the occupants.

This announcement tells on-scene crews that Command will not be at the command post and where he can be found if needed.

Reasons for Going Mobile

Among the reasons for Command's going mobile are the following:
• to talk to the occupants of the structure,
• to check on the status of injured civilians or personnel,
• to begin origin and cause determination, and
• to quickly check the status of exterior fire areas.[2]
Once Command returns to the command post, he should announce:

Noble Street Command is back at the command post.

TYPES OF COMMAND POSTS

The three types of command posts—mobile, stationary, and formal—vary in location, size, and complexity, according to the following:
• the type of command established,
• the needs of the incident at the time it is established (some "small" incidents, such as a small haz-mat spill in the downtown area at noon on a Wednesday in August, may have numerous needs), and
• the potential for the incident to expand (it's better to keep looking ahead than always to be trying to catch up).

Mobile Command Posts

With the mobile command post, it is expected that Command will be in a location remote from his vehicle but one from which it will be advantageous to focus on the whole incident. At a garage fire, the engine may be on the street or in the driveway, but the command post may be in the backyard. At a vehicular accident, the command post may be several hundred feet ahead of the first-in engine due to traffic tie-ups. The key characteristics of this type of command post are that there is no specific location for a command post and that the incident commander will be remote from his vehicle or other more traditional forms of command posts. When using a mobile command post, Command must announce its location if it is not obvious. A mobile command post is used for small incidents.

Stationary Command Posts

The stationary command post is used at normal single-family and multi-family dwelling fires and small commercial-structure fires. The Toledo Fire Department has instituted a procedure that designates the area around the first-in engine as the location of the stationary command post at these types of incidents. Unless Command designates another area for the command post, that is where Command is expected to be. When I arrive at a working incident and want to take command from the first-in officer, I automatically

Figure 4-2. Fire officer working at a mobile command post. (Photo by author.)

report to the first-in engine.

After a brief exchange of information, I usually assume command. If at that time I feel it would be more advantageous to move the command post to another location, I announce over the radio that I am changing the location of the command post. The location may be changed for many reasons, including the following:

- The new location may give a better view of at least two sides of the structure.
- The original post may be within a smoky or hazardous atmosphere.
- Command will be better able to step back and focus on the whole incident from the new location.
- The new location is a more logical site for other arrving members and outside entities to locate.

Stationary command posts tend to be less formal and are intended to be visited by a minimum number of entities (such as police command, the gas company, and other chief officers).

When using the stationary command post, it is more likely that Command may go mobile and leave the post unattended for short periods of time.

Stationary command posts usually are used at incidents where formal command is established and the needs of the incident are relatively small. Such incidents would include the normal residential structure fires and larger multivehicular traffic accidents.

Figure 4-3. Chief officer operating at a stationary command post. Note his proximity to the first-in engine. From this vantage point, he can focus on the entire incident.

THE FORMAL COMMAND POST

The formal command post, a necessity at multiple-alarm incidents, is more structured than the stationary command post and should be functional. It is used at large incidents where formal or unified command has been established. Its location must be formally announced over the radio. It is expected that the formal command post will always be occupied at least by Command. The dispatcher should announce the location of the formal command post as a reminder to units responding and listening. The tools necessary for a formal command post are discussed later in this chapter. The formal command post is used at large incidents where formal or unified command has been established.

Figure 4-4. Chief officers operating at a formal command post.

THE TYPE OF COMMAND POST AND THE TYPE OF COMMAND

Informal Command and the Command Post

Since an incident managed under informal command is usually small, its scope and needs are small. Traditionally, the unit responding may need to park on the street and walk some distance before the actual victim or fire can be seen. In these instances, Command would be mobile. The needs of these incidents are small, and, from the standpoint of focusing on all the incident's specific needs, it would be best for the individual in charge to go to the incident area instead of remaining in the street—hence, it is not necessary to establish a formal command post. Command will go with his crew to the incident if it is remote from the responding apparatus. The officer should make this fact known as part of his on-scene report.

Vehicle fire: Engine 7 is a 545 N. Huron. We have a car totally involved with no exposures.

OR

EMS run: Engine 7 is at 1234 South.

Table 4-1. COMMAND MODE VS. TYPE OF COMMAND POST

Command Post Type	Type of Command	Advantages/Characteristics
Mobile	Informal or formal	Allows Command to focus on the incident's needs. Incident may be some distance from apparatus.
Stationary	Formal	Allows Command to be in a recognizable position and still view the incident's needs, normally, at first-in unit (engine). Command can still move about or go mobile if necessary.
Formal	Formal or unified	Stationary and formal. Forces Command to stay at the command post. Others can "move about" for command. Should be recognizable.

Once on the scene, Command should direct his crew to handle the situation while he ensures that all the needs of the incident (i.e., requesting an ambulance, notifying police, securing the scene/house if victim is to be transported, and calming family members) are being addressed. Any or all of these actions can be delegated to crew members if numerous tasks must be accomplished.

Formal Command and the Command Post

Under formal command, more than one officer responds. One individual establishes command and directs other incoming units. A command post, which must be stationary, is established. The incident probably will be larger, and the concurrent needs of the incident will be greater.

Unified Command and the Command Post

A large, stationary formal command post—under a roof, if possible, and with appropriate lighting and communication facilities—must be established under unified command. Command will not be expected to go mobile. This incident will be of longer duration, and the numerous uniformed and civilian individuals responding will want to make their presence known at the command post.

Table 4-2. COMMAND POST TYPE, LOCATION, AND COMMAND TYPE

Command Post Type	Normal Location of Command Post	Command Type
Mobile	Logical area from which Command can focus on the incident, in a house or on the street	Informal; sometimes formal
Stationary	In front of structure by the first-due engine at the scene or in a parking lot	Formal
Formal	At a street intersection or in a parking lot	Formal or unified

DRIVE-BYS

At most working incidents, Command is required to remain at the command post as much as possible. Chief officers used to roaming around the fire find it hard to do this. Command must resist the temptation to continually go mobile. Previously assigned sector officers and responding units will want Command to be available for face-to-face discussions, updates, and sensitive notifications. They shouldn't have to circle the fire building trying to catch up with Command. The officer in command must learn to rely on sector officers' observations regarding out-of-sight activities and updates.

To initially observe as much of the fire building as possible, it may be beneficial for Command to do an intentional drive-by of the structure. By pulling up to and then past the structure, Command normally can see three of the four sides of a building. I have seen chiefs intentionally drive around the block to see all four sides of the structure prior to parking and setting up a stationary or formal command post. This extra 30 to 45 seconds may be more effectively spent creating a to-do list and then making initial assignments.

Figure 4-5. Drive-by shot of Side 4 of the structure. (Photo by author.)

Figure 4-6. Drive-by shot of Side 1 of the structure. (Photo by author.)

Figure 4-7. Drive-by shot of Side 2 of the structure. (Photo by author.)

COMMAND POST LOCATION

Apart from what already has been said with regard to the location of the command post, I offer the following rules of thumb:

- The command post should be in a location advantageous for focusing on the whole incident.
- The larger the incident, the farther away the command post should be.
- At the "Big One," it may be advisable for Command (and the command post) to be so remote from the incident that Command cannot view the structure involved (such as in the case of a high-rise fire). The National Fire Academy runs scenarios that prove this point quite vividly.

Figure 4-8. Stationary command post located relatively close to othe incident so Command can focus on the needs of the incident.

Figure 4-9. Large incidents require a formal command post. It should be far enough away from the incident to allow Command a larger focus area.

BASE OF OPERATIONS FOR COMMAND STAFF

Who should operate at the command post and who need not be there are up to Command's discretion. To be sure, Command needs to be there.

In general, Command will want most of his staff at the command post (if staff members are designated). The specific members and their functions will be discussed in the next two chapters. For convenience, however, following are the locations from which the most commonly used command staff members should operate:

- Command—command post.
- Liaison—command post.
- Safety—wherever the incident dictates, but normally not at the command post.

Figure 4-10. Setting up a formal command post at a large incident.

- Public information officer—at a location remote from, but within walking distance of, the command post.
- Operations—command post.
- Planning—command post.
- Logistics—usually the command post but may have to leave for short periods.
- Administration/Finance—command post.

Endnotes

1. Brunacini, Alan V. *Fire Command*. 1985. NFPA Publications.
2. Under normal circumstances, Command should *never* enter a fire building until the fire is deemed under control and, as a rule, only when SCBAs are no longer required. If Command needs information, he should rely on the interior sector officers for updates.

REVIEW QUESTIONS

1. Define "command post."

2. Who determines the location of the command post?

3. Define "mobile command."

4. What are some reasons for Command's "going mobile"?

5. Where would the command post usually be when operating under informal command?

6. What are the three types of command posts?

7. What factors determine which of the three basic types of command posts is used?

8. What are the reasons for changing the location of the command post?

9. Define "drive-bys."

10. What is one of the rules of thumb for choosing the location of the command post?

DISCUSSION QUESTIONS

1. According to the author, Command should remain at the command post for the majority of the incident. Discuss incidents to which you responded when this rule was not followed. Brainstorm and come up with a few reasons in addition to those given in the text that may cause Command to leave the command post.

2. Discuss the pros and cons of having Command and incident staff members operate from the locations given at the end of the chapter. In your opinion, is this list too rigid, or should it be more rigid in defining where the key players are expected to best focus on their assigned tasks?

3. Describe the type of formal command post your department uses. What are its good points and the areas that could be improved? Several departments have large command post vehicles. Compare and contrast them with what you presently use. Attempt to justify the expenditure for a large command post.

The Command Staff

TOPICS

- **Command staff positions**
- **The safety officer's role**
- **The relationship between Command and the safety officer**
- **The role of the liaison officer**
- **The relationship between Command and the liaison officer**
- **The role of the public information officer**
- **The relationship between Command and the public information officer**
- **Where the three command staff officers should operate**

Except for the tactical evolutions, Command can handle the needs of the vast majority of incidents to which we repond. Command should assign companies or personnel to handle the specific tactical evolutions that will stop the problem. While sector officers manage these tactical evolutions, Command can focus on the other specifics of the incident. Command must ensure that the members operating at the scene are as safe as possible and that all the appropriate safety equipment is being used. In addition, Command handles any concerns that outside agencies, such as the gas or electric company, may have relevant to the incident. Police and other response agencies will need to find where they can fit into the incident. Finally, Command must ensure that the public receives correct, up-to-date information. If the media asks for information concerning the incident, Command is required to provide answers and uphold

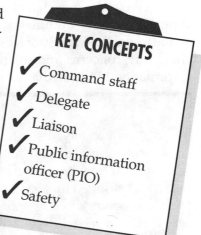

KEY CONCEPTS

✓ Command staff

✓ Delegate

✓ Liaison

✓ Public information officer (PIO)

✓ Safety

the proper image of the department. However, some incidents will be best served if Command delegates some or all of these responsibilities.

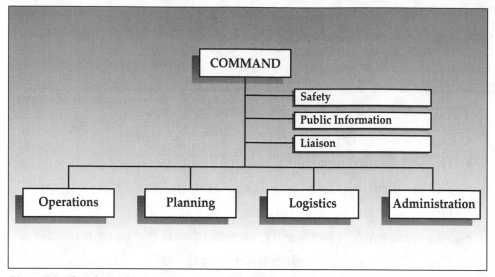

Figure 5-1. The relationship between Command and his staff.

THE PREDETERMINED COMMAND ROLES

Command must fill three predetermined roles at every incident: safety, liaison, and public information officer (PIO).[1] He has the option of assuming these roles or delegating them to other officers at the scene.

As seen in the flowchart (Figure 5-1), command staff officers, if assigned, report directly to Command. Even at the smallest incident (informal command), someone is responsible for the safety of the crew and civilians, handling any concerns and interactions with outside agencies, and providing information to the media (if asked).

SCENARIO 1

Let's say you are an engine officer assigned to an inner-city company. At rush hour, you're dispatched to a car fire on Twelfth at Madison Street. On arrival, you find a passenger van totally involved. The van is sitting off the road and stopped up against an electric company pole. There are no structural exposures, but the pole and an electrical drop to a convenience store are involved. The driver of the van is outside and unharmed but tells you that he was involved in a hit-and-run, the force of the collision ruptured his gas tank, and he isn't sure about how the fire started. You tell your crew to mask up and attack the fire (while out of the corner of your eye watching the rookie to make sure he remembers to put his hood on). Next, you call for the electric company and a police crew for traffic control and to file a report. As your crew is working on the fire, the police arrive. You tell them what you need and then notice that a news crew has arrived. About this time, the electric company "trouble" crew arrives. You discuss the situation with the lineman from the electric company, who says that if it's okay with you, he will pull the line clear until you're done and repair it after you leave. You tell him that's fine. A few minutes later, a reporter asks you if you have the time to tell him what happened.

At the above incident, you were the safety officer, liaison, and PIO. You handled the incident and still filled those three positions. At 99 percent of the incidents to which I respond as a battalion chief, that's how it works. But every once in a while, I can't meet the needs of the incident and still fulfill those positions as well as they should be, so I delegate.

Figure 5-2. At incidents such as this, Command probably can fill the roles of safety, liaison, and PIO and still meet all the needs of the incident. (Photo by Ken Anello.)

The Role of the Safety Officer

Command is ultimately responsible for the outcome of the incident and, as such, for the safety of response personnel and civilians at the scene. However, Command must make every attempt to remain focused on the whole incident. If this means that he must assign a specific safety officer, then so be it. At most single-family structure fires (our "bread and butter" fires), Command can probably do both—focus on the whole incident while addressing the safety aspects of the incident.

Note: Some of the responsibility for safety resets with the company (sector) officers. More will be said about this is Chapter 8.[2]

Responsibilities of the Safety Officer

The safety officer is responsible for scene safety. If Command assigns a safety officer, that officer's radio designation most likely would be "Safety." Command must work on the assumption that the officers are doing what a *reasonable, prudent* officer would be doing in the same situation. Safety should go where the most problematic areas are for that particular incident.

Since Command has a priority list for the incident, Safety should have a priority list for the safety concerns relating to that particular incident. At one fire, those concerns might be the following:

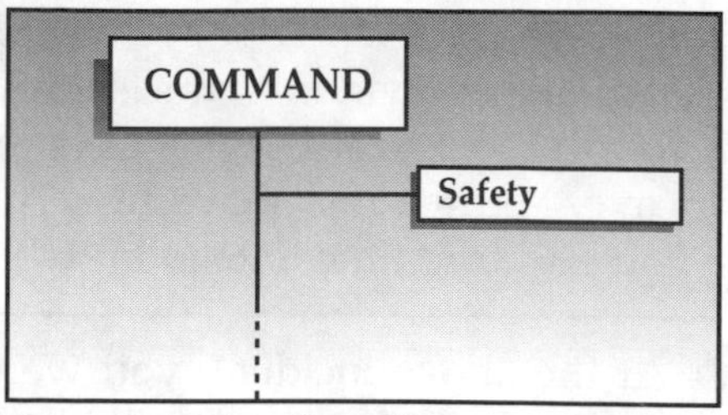

Figure 5-3. The safety officer.

- electrical wires down across the front yard,
- ladder operations for roof ventilation,
- the interior heat conditions of the structure at or near the attic, and
- the safety of the civilian bystanders.

At another incident, the safety concerns might be totally different:

- the proximity of flammable liquids to a rapidly moving fire,
- the potential for explosion as it relates to civilian bystanders, and
- the safety of the ladder company attempting a trench cut on the roof.

Safety develops the list alone or after an initial consultation with Command. However, after the initial consultation, Safety should move from one area to the next, ensuring the safety of that particular area. Once the safety concerns of that area have been met, he can move on to the next area.

Who's the Boss

An age-old question is, What if the safety officer sees an unsafe act ordered by Command or Operations?

Figure 5-4. The safety officer should be mobile so he can be at the incident's most problematic areas. Notice the proximity of the electrical wires to the aerial.

Figure 5-5. The safety officer needs to roam the scene. Notice the electrical drop on the auto's roof and the hoseline draping over the same auto. Safety needs to focus on these types of scenes.

Let's say Command ordered that the roof be opened. Safety, hearing this, decides to go up the ground ladder and notices that the roof is beginning to sag and that smoke is forcibly coming out between roofing shingles.

In my opinion, Safety should stop the activity. He should order members off the roof (or onto the ground ladder) and simultaneously inform Command of the situation. After that, it's up to Command, who has formally been put on notice that a hazardous condition exists. To overrule the safety officer and order crews back on the roof would not be prudent. If members were to go back and then get injured or killed, Command could be held liable for their injuries or deaths. (Let's face it! If this were the case, the roof would vent itself momentarily anyway. To risk crews for what Nature will accomplish in a matter of minutes is not only unnecessary but a waste of resources.)

Training

In my opinion, any officer who may be designated as Safety should have, as a minimum, prior training in the following:

- firefighter safety practices,
- building construction,
- haz-mat operations, and
- air monitoring and confined-space operations.

If chief officers are to function without trained safety officers, then they should be trained at least in the above areas.

This section is not intended to be all-inclusive with regard to the safety officer's operations and responsibilities. NFPA 1500, *Standard on Fire Department Occupational Safety and Health Program*, and 1521, *Fire Department Safety Officer*, meet that need quite well. My intent is to show the relationship of the safety officer to Command at an incident.

The Role of the Liaison Officer

The liaison officer is the office manager of the command post. He determines who can (and should) have a face-to-face conversation with Command. He is the member of Command's staff who will meet with outside agencies and channel them to where they can best assist in the incident. A liaison officer could be a lifesaver for Command at a large incident.

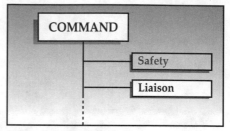

Figure 5-6. The liaison officer.

Responsibilities of the Liaison Officer

* The liaison officer has three basic responsibilities:
* to greet persons reporting to the command post,
* to serve as a mouthpiece for Command. There are times when Command is too busy to give directions to individuals reporting to the command post. The liaison officer can pass messages and give directions at these times; and
* to direct outside agencies to where they can best fit the incident's needs. The gas or electric company, for example, can be given a brief direction and then sent on its way; the police can be told which streets need to be blocked; and so on. All of this frees up time for Command to "look important."

Who Should Be the Liaison Officer?

The person assigned as liaison officer at a large incident should be someone who can sort out the complexities of an incident and get the most out of the people who report to the command post. Liaison should be experienced at firefighting strategies and logistical operations and should be able to work well with others (including individuals from fields other than firefighting). This is the perfect place for "senior" members of volunteer departments who have a wealth of knowledge but have a hard time moving around the fireground.

Where Should the Liaison Officer Operate?

The liaison officer should work at the command post with Command and should not be involved in most of the strategic discussions concerning the incident. Liaison's sole task is to greet individuals from other city, county, safety, or outside agencies such as utility companies, and be Command's mouthpiece. After briefings with Command and other command post personnel, the liaison officer greets outside agency members allowed to report to the command post. At large incidents, law enforcement officers—police, sheriff, state patrol, or military police—should secure the perimeter of the command post area to sift and channel people approaching the area.

I have been at incidents where intoxicated "do-gooders" almost caused riots in the command post area because Command could not or would not take the time to listen to their suggestions. At large incidents, a perimeter should be established at the command post with scene tape, and entry should be permitted only at the request of Command or the liaison officer.

As stated above, the task of the liaison officer is to meet with outside agencies reporting to the incident. Freelancing by personnel *or* outside agencies should never be tolerated. In Toledo, our dispatchers have a "call-out" sheet that they reference whenever we have a second-alarm or greater incident. The list contains the names of specific agencies and their representatives who should be or would like to be notified of a large incident. Included are the following:
- utility companies—electric, gas, phone, cable, and so on;
- hospital emergency rooms;
- higher ranking chief officers within the department; and
- police command.

When notified, the individuals who choose to report to the incident should report to the command post. If the incident requires that Command's attention be directed in other areas, he should designate a liaison officer, who will greet these individuals and explain their roles in the incident. At times, individuals with personal or economic interests in the incident will report to the scene. They will have to be dealt with also. If the individual securing the area around the command post is not sure whether a person should be allowed access to the command post, then Liaison can be contacted and a decison made.

The liaison officer should not act without direction from Command, and Command should make clear what he would like from specific agencies expected to report to the incident.

Finally, the liaison officer can be very helpful in placating elected officials who report to the incident. He can act as a buffer and redirect these well-intentioned individuals to areas where they can expend their energy in a more useful manner. Seriously, we will not stop elected officials from coming to large incidents. I have found that if grouped together and then periodically given a report on the incident, elected officials usually are not a problem. The key is to do your best to make them feel important and to feed them information in a timely manner. One firefighter or officer assigned to these folks can save a lot of on-the-spot frustration or next-day explaining. An experienced liaison officer can save Command from many distractions.

The Role of the Public Information Officer

The PIO is the member of Command's staff charged with interacting with the news media. In many ways, the PIO can make or break an incident for Command. What (and even how and when to some extent) he says to the press can make Command (and the department) look important or incompetent. It must be remembered that if you are Command and if you designate someone else as PIO, that person will be speaking for you. Choose your PIO with care.

Responsibilities of the PIO

The PIO's responsibilities include the following:

- to periodically contact Command or Operations for updates and information.
- to provide information to the media concerning the incident.
- to provide information to other outside agencies and officials.
- to provide information to civilians who have a vested interest in the incident. This would include, for example, civilians who have been evacuated due to a fire or haz-mat incident.
- to provide information to incoming crews or other agencies operating at an incident. It may be best if the PIO conducted scheduled briefings on the situation.

The PIO should operate at a location slightly remote from the command post. There are two reasons for this. First, it keeps the ears of the media away from some of those sensitive conversations that take place at the command post—such as the final decision to "let the building go!" or the fact that Search found a victim on the third floor and has chosen to leave the body there. We can certainly explain either of these statements, but we might want to do it in a prepared statement as opposed to answering a barrage of questions resulting from someone's overhearing only part of a conversation. Second, it removes the inevitable distraction of having a camera or microphone in Command's face.

At most of the fires I command, I enjoy being my own PIO. I have two children at home who love seeing their dad on TV once or twice a month. I also enjoy having people say, "Saw you on the news last night." I have had a few run-ins and have been misquoted a few times; but for the most part, I enjoy providing info to the media. However, there are times when it just isn't practical for me to stop and talk to them.

I recall a fatal second-alarm fire in an apartment building. Just about the time the news crews were pulling the cameras from the box, Chief Michael P. Bell arrived on the scene and asked me if I needed anything. I told him it would be great if he would be the PIO and let me finish the fire. (At that time, I was heavily into searches, PPV, and the coroner's office and didn't need mikes in my face.) The chief took the media all the way to the back of this big parking lot. He designated a "runner" to go back and forth with information. This worked out very well. He took those people out of my arena and still kept

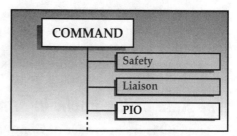

Figure 5-7. The public information officer (PIO).

Figure 5-8. A chief officer providing information to the media.

them satisfied. That is what a PIO can do. Every once in awhile, when it counts, Command should take this area of concern (when what is said may make a difference in public opinion) off of his own shoulders.

Choosing a PIO

Some thought should be put into choosing the PIO. This individual should be knowledgeable in firefighting practices and able to explain the "whys" of the incident's observable operations in terms that help the public image. The PIO should be someone who displays a good positive image for the department.

Normally, I try to delegate the responsibility of PIO to another responding chief officer. At a large incident, several are usually available. Staff deputy chiefs or the chief/director of the department normally have had considerable experience in front of the camera and are well-equipped to answer even the most stressful questions.

The liaison, PIO, and safety officer positions are tools in the incident management system that are available if needed. They are there to make Command's job easier. If the needs of the incident are being met and Command and the incident can afford it, then Command can handle these three positions. You must remember one thing: If Command—regardless of whether the management structure type is informal, formal, or unified—does not delegate these responsibilities to others, then Command is responsible for them.

Endnotes

1. The National Fire Academy refers to this staff position as the "information officer" because this individual can also give updates and information to uniformed employees such as incoming crews; safety forces; and other city, state, and federal agencies. "PIO" would indicate giving information only to civilians and the public. I prefer to keep the position as PIO. Command can require that this individual give information to any number of agencies. Although I understand the reason for the change, I feel it is not the time to change the name.

2. Command must work on the assumption that his officers are doing what a reasonable, prudent officer would be doing in the same situation. Command cannot be everywhere, nor can a safety officer. You generally cannot be held accountable for unsafe acts committed unless you know they are being committed.

REVIEW QUESTIONS

1. Define "command staff."

2. What are the three command staff roles?

3. Command has the option to perform or _____ these roles.

4. Some of the responsibility for the crews' safety must rest with whom?

5. What is the safety officer's responsibility?

6. How should the safety officer create his to-do list?

7. To gain optimum effectiveness in scene safety, the safety officer must be _____.

8. The safety officer should look out for the same safety concerns at every incident. True ___ False ___

9. List one of the areas in which Safety should have prior training.

10. The author refers to the liaison officer as the _____ of the command post.

11. List the three responsibilities of the liaison officer.

12. List an entity that should report to the liaison officer.

13. What is the role of the PIO?

14. Where should the PIO operate?

15. Name one individual who can be assigned to be the PIO.

DISCUSSION QUESTIONS

1. Discuss in your own words the concept of "reasonable, prudent" actions taken by Command and Safety. Explain how it relates to the text.

2. What are the four areas in which the safety officer needs at least additional training? Create your own list of areas of training that a safety officer may find worthwhile.

3. Assume you are the chief of a small community fire department. You respond to a large incident in your community. You call for mutual aid and have three other department chief officers responding. One is a young, aggressive, and progressive chief; another is an older veteran chief who has allowed the department to remain unchanged (in the area of innovation and acceptance of new concepts) for the past 15 years. The last chief is old, experienced, and very progressive. Each chief will be required to fill a staff role. Based on what you know, assign each chief to the role for which he is best suited.

The Functions of Command

TOPICS

- **Functions of Command**
- **Role of the operations officer**
- **Operation's focus at various types of incidents**
- **Role of Planning**
- **Responsibilities of Planning**
- **Role of Logistics**
- **Responsibilities of Logistics**
- **Role of Administration**
- **Responsibilities of Adminstration**
- **Where section officers operate**

KEY CONCEPTS
- ✔ Accountability system
- ✔ Administration section
- ✔ Defusing
- ✔ Documentation
- ✔ Logistics section
- ✔ Operations section
- ✔ Planning section
- ✔ Procurement

Command must perform four functions—referred to as the "four functions of Command"—at every incident, regardless of the incident's size or complexity. These functions often are confused with the four priorities of Command, covered in Chapter 2 (firefighter safety, civilian safety, stop the problem, and conserve property). These priorities are criteria that Command should consider when making strategic decisions and assignments at incidents. The four functions of Command are quite different. They can be considered areas of responsibility that Command *must*, in some fashion, fulfill at every incident. These functions have nothing to do with to-do lists or sector assignments. They are more administrative or managerial in scope. If Command can manage these four functions along with the responsibilities of his staff and meet the needs of the incident, then Command probably does not need to delegate any of these functions. However, if Command is becoming bogged down with minor tasks, some or all of these functions should be passed off to others.

Functions are also referred to as "sections," which are discussed later in this chapter.

FUNCTIONS OF COMMAND

The four functions of Command are the following:
- Operations: Managing the strategic aspects of the incident.

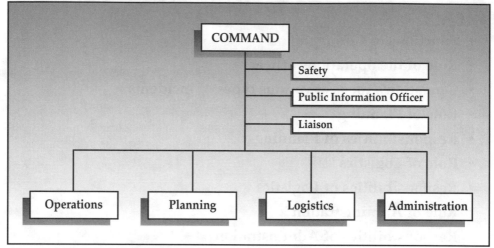

Figure 6-1. The four functions of Command.

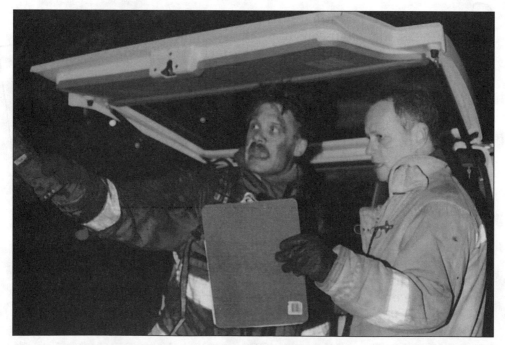

Figure 6-2. Command discussing options with an officer. This incident occurred in a vacant industrial heat-treating plant with abandoned hazardous materials.

- *Planning*: Reviewing the past and preparing for the incident's current and future needs.
- *Logistics:* Bringing the required tools and equipment to the scene.
- *Administration/Finance:* Ensuring that the administrative aspects of the incident are being met.

As mentioned, all of these functions must be fulfilled at every incident. They are things officers have been doing at every incident, but now these tasks have been given names and have had their roles formalized and their existence justified.

SCENARIO 1

I leave my station in the morning for headquarters. At a gas station on Dearborn and Front Streets, I notice a car fire in the parking lot.

Battalion 1 to Dispatch: I have a car fire at Front and Dearborn. Send me an engine company. Battalion 1 is Front Street Command.

I pull up in front of the car, get out, and put on my turnouts. I grab an extinguisher and begin to hit the car fire. At this point, I am Operations. (I can strategically handle this incident. I have additional units coming for backup.) I can also be Planning. (There are no hazardous materials involved. There are no serious exposures. I can manage and account for the resource status and how to demobilize the incident when it's over.) I can also handle logistics. (I have all the tools necessary for the fire. It's a small engine fire and my multipurpose extinguisher should do the trick.) Finally, I can handle the administrative aspects. (I will not need any legal advice, and there will be no overtime at this incident. We will refill the extinguisher at a later time, and the bill will be absorbed by the department. There are no injuries—I hope!—so no workers' compensation claims need to be processed.)

The above example is an oversimplification of the four functions of Command. However, the example is given to illustrate that this is nothing new. Now, let me take this example a step further.

SCENARIO EXTENSION

I leave the car fire and continue downtown. As I approach headquarters, Dispatch tells me to "stand by for a structure fire." I am dispatched to 711 Locust on an apartment fire. I am the first to arrive at 711 Locust, where I observe light smoke coming from the second floor of what appears to be a vacant apartment building.

Battalion 1 is at 711 Locust. We have a working fire in a three-story apartment. Battalion 1 will be Locust Command.

I assign Engine 3 as Division 2 attack, Engine 5 as backup, Squad 1 as search, and Truck 13 as ventilation. *(I have fulfilled the functions of operations.)*

These companies will enter the structure and complete the tasks assigned them. I can handle the function of planning at this incident. There are no exposures. I can account for all members assigned and keep updated on the situation (with feedback from my officers). I can develop a plan for sending companies back when the incident is over. I can handle the logistics for this incident. I have all the tools and equipment I need. Air bottles will have to be exchanged (with an air wagon), and Engine 5 will need fuel after it leaves. Finally, I can handle the function of administration. There is no overtime, and there are no other administrative needs at this incident.

Again, I have handled all of the four functions of Command at this incident. As a last example:

EXAMPLE EXTENSION

I receive a report that several homeless families are living in this building. Attack reports that the fire has gotten into the walls and appears to be extending upward.

Battalion 1 to Dispatch: Give me a second alarm on Locust.

As my second-alarm battalion chief arrives, I become operations officer, and he takes command. (The reason for this move is explained later.) Now, Command has delegated the responsibility of operations to me. He will be responsible for planning, logistics, and administration at this incident.

The incident management system (IMS) allows for an easy flow of responsibility at incidents as they expand or shrink in size and complexity. Throughout, one person retains control (Command) over the entire incident and ensures that all the bases are being covered. These functions are often referred to as "sections"—for example, "Have one firefighter report to the logistics section to help with refueling vehicles."

Note: At the end of this text are some incident work sheets that can be developed for each of the functions of Command and his staff. These sheets are offered as suggestions. For most of us, multiple-alarm incidents happen only occasionally. To memorize all of the responsibilities involved would be ill-advised. Work sheets serve as reminders. All staff and off-duty chief officers and line chief cars should carry them. When functions and staff positions are assigned, the work sheets serve as refreshers.

My suggestion is that if you currently do not use incident work sheets or if they need updating, review what I have provided. Ask other departments for a copy of their work sheets and then develop one to fit your department's needs.

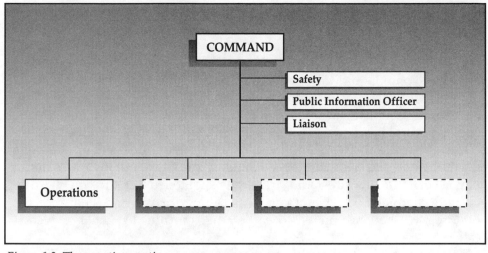

Figure 6-3. The operations section.

THE ROLE OF THE OPERATIONS OFFICER

The operations officer is responsible for managing the strategic aspects of the incident and for the following:

- Strategic decisions concerning the incident (after consultation with Command).
- Assignments (fire, EMS, or mass-casualty) of line crews. In an incident requiring fire and EMS services, Operations is responsible for managing both areas of the incident.
- Keeping Command informed. Command will be working at the command post with Operations, but Command may (and should) be occupied with other aspects of the incident. If developments and strategies change, then Command must be informed. The intent here is not that Command micromanage Operations. (If Command does that, then Command should take over operations.) Since Command is responsible for the outcome of the incident, however, Command must be kept informed of developments.

In Toledo, procedures state that an operations officer must be assigned at any second-alarm or larger incident, or whenever Command deems it necessary. In addition, our procedures provide that when the second-alarm chief or a higher-ranking chief arrives, the initial Command move to the position of operations section chief and the next-arriving chief serve as Command. This procedure, established after the following incident, works very well.

THE INCIDENT

One Saturday morning, I was dispatched to a vacant industrial structure fire in the heart of the city. Shortly after arrival, I called for a second alarm. Battalion Chief Joe Walter arrived, and I designated him Operations. At that time, I had established a command post at the rear of my Jeep® and had

the command board out. It took me about five minutes (and many inter-
ruptions) to explain to him what I had done, my strategy, and who was
doing what—and where. After the fire was out, we were talking and cri-
tiquing what we had just gone through. He said it might have been easier
if he had come in and taken command and let me move down to operations. I had had a specific strategy in mind and already knew what assignments had been made. It sure would have saved a lot of explaining. We discussed several other aspects of the incident, and he finally left the scene.

I left the scene about an hour later. We kept three companies on the scene
for fire watch and overhaul. About an hour later, I was in my office doing
paperwork. On the fire radio, I heard "heightened" chatter on the fireground channel. Soon, Command was calling for a regular alarm at the
building we had responded to earlier that morning. When I arrived, I
called for a second alarm.

This time, when Chief Walter arrived, he took command; and I went
down to operations. This time, the transition and the fire went much more
smoothly. There was little lag in the transfer of command. Chief Walter
came in and began to concentrate on planning, logistics, and administration. I was left with operations. No time was lost explaining what I already
had in place.

Radio Channel

Whenever an operations officer is established (or any other time at the
direction of Command), Command should go to a separate radio channel, if
one is available, for the following reasons:

Figure 6-4. The operations chief discussing strategies and expectations with a chief assigned to Sector 3.

Figure 6-5. At the second second-alarm fire in the same building on the same day, the original Command moved down to operations to save time and confusion. This fire went much better.

- There is no need for Command to listen to conversations such as an engine officer's calling for "20 pounds more pressure on the line" or "Hey, George, swing that aerial about 10 feet to the left." That presents too much of a temptation to micromanage. If Command has assigned a competent operations officer, then 100 percent of the fireground chatter is unnecessary to Command. (If something goes wrong, Operations will let Command know. Keep in mind that they should be standing right next to each other.)
- Command shouldn't have to worry about cutting off fireground discussion among units assigned to operations when he wants to talk to dispatchers about getting Red Cross assistance or checking to see if the mayor has been notified.

Who Should Be Operations

The beauty of the IMS is that it is truly user-friendly. It works with departments of all sizes and at incidents of all types. Toledo's procedure that, at a large incident, the second-arriving chief officer take over as Command works in larger departments, where chief officers from the same department will arrive.

But what about smaller departments that rely on mutual aid for their large incidents? It may be more prudent for these departments to establish a procedure that has the original chief (home chief, if you will) retain command. The first-in mutual-aid chief and Command should take a few minutes to review original strategies and assignments, and then Command should assign Operations to the mutual-aid chief. The "home chief" may be better equipped to handle the other functions of Command (i.e., planning, logistics, and administration). He would have a better grasp of the structure and its complexities (planning), the resources available in the town or jurisdiction and neighboring communities (logistics), and the procedures and costs associated with the incident (administration). Any responding chief officer, after a short briefing, should be able to run the incident strategically—especially if a uniform system of staging and sectorization has been established.

The operations officer should be well versed in the strategic concepts of firefighting, emergency medical operations (specifically mass-casualty management), and haz-mat operations. Senior-level officers with past experience should be chosen over less experienced officers. (As you will see, there are plenty of assignments to go around.) In some instances, the position of operations officer could be used as a training ground for younger officers. With a strong Command at his side, a "rookie" officer could be "walked" through a vacant house or structure fire while the more experienced chief officer, as Command, looks on. These occasions may be rare, but opportunities do present themselves at times and should be used to advantage.

In rural jurisdictions, controlled burns, such as organized barn fires, provide training operations for potential operations officers.

Focus of the Operations Officer

At a fire, Operations should make the larger, strategic decisions, including the locations for launching interior or exterior attacks and the type of ventilation to be used. Operations is responsible for ensuring that the appropriate number of backup lines be assigned. Operations should not, however, micromanage specific tactical evolutions or get involved in the specific fire attack, the nozzle application, and overhaul tactics. Once Operation's focus turns to specific tactics, he will lose sight of the overall incident action plan. Everyone on the fireground—including staged units (their assignment is to stage)—has an assignment, and they should be focusing on their specific assignments. Once focus is lost or those assigned to a task are distracted by other tasks, the original objective becomes blurred or lost. That's the time injuries occur or buildings are lost. Operations must focus on the strategic action plan of the incident. Sector officers should employ specific tactics to achieve their goals. More will be said about this later.

At a mass-casualty incident, Operations should focus on ensuring the extrication, triage, treatment, and transportation of victims. The majority of these tasks will be delegated to other on-scene officers. An officer on a heavy squad may be put in charge of extrication. A medic may be in charge of triage, treatment, and transport. Operation's role is to ensure for Command that the needs of victims and emergency crews are being met. Command should focus on all the other scene functions (or delegate them to others). These may include planning any haz-mat needs associated with the incident, getting the necessary cutting tools to the incident, and setting up necessary field hospitals or procuring air ambulances.

At hazardous-materials incidents, Operations normally attends to three phases of the incident: fire/backup, haz-mat, and emergency medical needs of victims. These phases are explained in Chapter 22.

Where Should Operations Operate?

Figuratively, Operations is sitting in Command's back pocket. They should both be at the command post and within talking distance of one another. Some textbooks say the only logical place for a chief officer during a fire is at the nozzle. This is not at all practical. As stated in Chapter 4, the larger the incident, the farther away the command post should be. Incidents requiring an operations officer are large. Command should be involved in major strategic decisions. The decision to go from offensive to defensive operations should be made only by Command, after consultation with the operations officer. The decision to move ventilation companies from a spongy roof should be discussed with Command. (If safety concerns prohibit the initial discussion, as soon as the crews are in a safe location, Command must and should be advised; and the final decision will rest with him.) The operations officer should decide where to place exterior exposure lines, assign search crews to perform secondary searches, and determine the priority of salvage operations. Command shouldn't have to worry about such minor decisions.

THE ROLE OF PLANNING

If the technical aspect of the incident begins to overwhelm Command or Operations, then it may be prudent for Command to designate a planning officer. The planning officer is responsible for reviewing past and then identifying current needs of the incident and may be required to anticipate the future course of the incident. Any technical areas that need to be addressed, such as hazardous-materials specialists, chemists, or transportation experts, will work with the planning section officer to provide Command with a view of the past, a realistic look at the present, and a glimpse into the future.

Planning normally is responsible for the following:
- Incident Planning
  —Evaluate and update the current strategic plan with Command and Operations.
  —Evaluate past actions and strategies.
  —Refine current and future plans and recommend any changes to Command and Operations.
  —Forecast possible outcomes.
  —Evaluate future resource needs with Operations.
  —Evaluate to-do lists established by Command and Operations.
- Resources Assessment
  —Maintain an accounting of on-scene personnel.

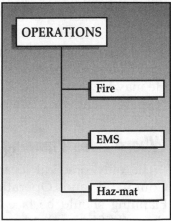

Figure 6-6. Sections under Operations.

—Maintain an accounting of the specific units dispatched/assigned.

—Maintain an accounting of special equipment on-scene and responding.

• Situation Status

—Maintain an up-to-date account of the incident.

• Documentation

—Maintain incident records (i.e., time lines and request times).

—Maintain current command chart; evaluate and maintain an effective span of control.

• Technical Specialist

—Coordinate with outside specialists such as chemists and HVAC engineers.

• Demobilization

—Create a plan for closing out the incident.

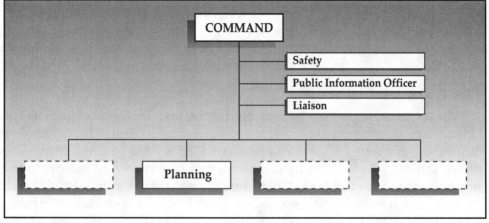

Figure 6-7. The planning section.

Who Should Be Assigned to Planning?

The planning officer should be able to work methodically alone or draw the best from individuals. He must have excellent organizational skills and be able to lead a group of people through a problem. As you can see, the planning officer must be able to work with a diversity of people. He must be a diplomat and a politician. At times, the planning officer will have to present "bad news" to Command or other section officers at the command post. He may have to "persuade" Command or others after consultation with the "experts" that what "we had hoped for will never be."

Where Does Planning Operate?

The only practical location for Planning is at the command post with Command and Operations. Conversations between Command and Planning should be face to face. Planning should monitor the command channel (if he needs to monitor any channel). Strategic and tactical information will come from Operations in the form of face-to-face discussions.

In-Depth Planning: Incident Planning

Fire Incident

At times, an incident can change almost minute by minute. The dynamics of these incidents make the tasks of Command and Operations difficult at best. At some incidents, the "unknown" is by far the biggest threat to an already unstable incident. On top of setting strategies and providing for the safety of on-scene crews, Command and Operations will have little time to just step back and think the incident through.

Among the reasons for this are the following:

- Operations is focusing on the strategic and tactical plan of the incident and should be receiving and evaluating information from sector officers and others on the fireground. He will be observing fire and smoke conditions and reviewing the command board to look for areas of weakness in the attack. Additionally, he will be receiving tactical information for relay to other sector or section officers.
- Command should be reviewing strategies with Operations, directing unassigned staff positions and functions, and monitoring the command channel for updates from dispatchers.

This leaves the above two individuals little time for looking back, thinking ahead, and discussing options with others at the scene. When the situation has more questions than answers and there is not enough time to sort between the two, Command should consider assigning a planning officer. (Remember, Command should *never* be too busy to talk with a subordinate or other officer at the command post. If Command is too busy to stop and talk, Command has too much to do.)

At structure fires, Planning may concentrate on the following:

- Construction type—its stability and effect on the fire.
- Occupancy type—its effect on the fire.
- Exposures—their construction and occupancy type and a priority list of exposure protection.
- Internal fire protection systems.
- Other internal systems such as elevators, HVAC systems, communications, and so on.
- Where the fire is going and where it can be stopped.
- Risk analysis—is what is being proposed worth the risk?

Haz-Mat Incident

- Material spilled and its properties.
- Safety concerns.
- Evacuation plans.
- Environmental concerns.

Mass-Casualty Incident

- Problems associated with cause and effect. What caused the problem and how will it affect rescues? If the incident is caused by a train derailment, where did it derail? What infrastructure is involved? What else was the train hauling? The list goes on.

- Prioritizing areas of rescue/risk analysis.
- Establishing alternate/additional treatment facilities.

Planning should reevaluate areas of concern during the incident and continually look at the picture and ask, What can happen next?

Resource Assessment/Situation Status and Documentation

Planning should coordinate with the safety officer to ensure an accountability system for all on-scene personnel. If outside agencies are also on-scene, a similar system of accountability must be established. This information can be cross-referenced with the information on Operation's command board.[1] If Command, Operations, or procedure calls for a Personnel Accountability Report (PAR), Planning should be the section that records the units reporting "PAR."[2]

To have a situation status (sit-stat) officer assigned under Planning is helpful when the situation warrants it. This officer would be responsible for maintaining an up-to-the-minute ac-

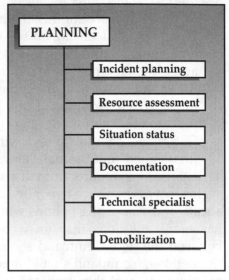

Figure 6-8. Sections under the direction of Planning.

count of the incident. This may sound like a waste of personnel, but, in a few instances, a sit-stat officer can save resources, time, and embarrassment. One firefighter sitting at the command post with the responsibility of writing down everything going on can be a lifesaver. Time checks can be established for the following:

- Notifications—coroner, haz-mat team, defusing team, benchmarks,[3] and so on.
- Calls for additional fire units or alarms.
- Calls for special equipment (dozers, endloaders, forklifts).
- Announcements of additional companies reporting to staging or special equipment arriving on the scene.
- Notification that outside agencies reported to the command post.

Again, some of this information may seem redundant. But you may have one incident in your career at which the price of keeping one more firefighter at the command post will save a life or save you from a lawsuit.

Technical Specialists

One of the problems with firefighting is that we usually cannot predict where fire or other disasters will strike. If we could, we would take measures to prevent or prepare better for the incident. Albert Einstein said, "A wise man does not have to know all the facts; he just knows where to find them." This is also true of fire chiefs and firefighters in general. Although we must know some fundamentals to perform at normal incidents, we cannot know everything.

Incident management recognizes this fact and builds in a function for the gathering of specific facts. If a chief, for example, is not too strong in the field of chemistry, should he put himself out of service if a haz-mat incident occurs? Should a chief know how a particular type of HVAC system will affect the spread of fire in a high-rise or hospital? No. But a chief should recognize the areas in which he is not knowledgeable and know how to gather experts in those fields. Meeting with college or industrial chemists and establishing a means for using their expertise may be a first step. Pagers are cheap. Some departments have agreements with specialists who provide their services for as little as a dollar a year. Fire companies should preplan commercial occupancies and list the building engineers and their assistants. They usually want to be called if an emergency arises within their facility. Who better to help answer questions about an HVAC system than the individual who works with it all year round? Scheduling some drills with these individuals will help keep the department sharp and make the facility's personnel aware of what to expect in a real emergency.

All specialists called to the emergency scene should report to the planning officer. With a little practice in the form of some small-scale disaster drills, these individuals will be able to help plan at the next large incident.

Demobilization

Other than logistics, demobilization is my biggest headache at a large incident. Demobilization's responsibility is to provide Planning (or Command) with a plan for picking up the incident: "Who can we send home first and who has to stay?"

The first thing a demobilization officer must do is get a paper and pencil and sketch the incident, plotting where the individual units are located. Operations already may have this information on his tactical command work sheet, and it's a place to start (Operations won't want to give that up, and it may be hard to carry that big board around a large fire scene). Next, Demobilization should ask Operations what types and quantities of equipment will be needed. After the sketch is completed and Operation's future plan of action has been obtained, Demobilization should take a walk around the emergency scene and determine which rigs have equipment on them, which have been stripped, which engines and trucks have lines (supply and attack) going into or out of them, and which units are just "sitting." That information will enable Demobilization to make a list of which units can go back first and which should stay. The list is then run past Operations, Planning (if established), and then Command.

THE ROLE OF LOGISTICS

Logistics is one of the biggest headaches at a large incident. For the most part, Command can handle procuring equipment, moving personnel, providing facilities, and other logistical needs at "bread and butter" fires. However, at a high-rise fire, where hundreds of SCBA bottles will be needed along with additional hose, nozzles, PPV fans, and so on, logistical needs can become quite complicated.

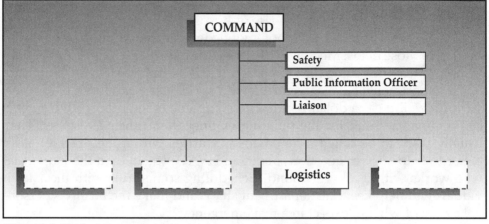

Figure 6-9. The logistics section.

Who Should Be Assigned to Be Logistics?

"Not me," you may be thinking. Some people have a gift for logically planning how to move things from one spot to another. These individuals are best suited to serve as logistics officers. The logistics officer need not necessarily possess the ability to determine what the incident's needs are (such as end loaders, SCBA bottles, welders, and so on). That is the job of Command, Operations, and Planning. Logistics must be able to fill requests for items as quickly as possible.

Rank should not be a factor in selecting a logistics officer. This individual should know what equipment is on hand in his department and surrounding departments and where to quickly get the items that aren't available. Additionally, the logistics officer should have a good understanding of how things work. While Logistics doesn't have to be a full-fledged mechanic or millwright, he should be mechanically capable and able to think ahead. Engines and aerials need fuel. People need fuel. SCBA bottles only last so long. Chief officers need the ability to talk to one another. The most effective logistics officer is one who can anticipate needs and efficiently provide the items needed when they are needed.

Figure 6-10 (above). Logistics would supply operators for the bobcat pictured here, which would be used to move burning grain at an elevator fire. Figure 6-11 (right). SCBA bottles need to be refilled. If Command does not establish a logistics section, then he is responsible for seeing that crews have full bottles.

Where Should Logistics Operate?

Logistics should have a home base at the command post. If the department has a command vehicle, a logistics (or resource) list should be carried in that vehicle. To be totally effective, however, the logistics officer may need to roam the scene to seek, disperse, and coordinate needed tools and equipment on the scene. If requests keep coming in for SCBA bottles and on-scene or stationary cascade systems can't keep up, Logistics may initiate a search of the scene for unusued bottles. When these activities are over or when responsibilities are delegated to others, Logistics should report back to the command post, which, as noted, should be Logistic's base of operations.

Responsibilities of Logistics

In some respects, Logistics is the place where the missing pieces are placed. Logistics ends up being a catchall for loose or unfamiliar responsibilities. Command needs to concentrate on the entire incident and make sure that all the bases are being covered. Operations and Planning will have their hands full focusing on their specific tasks. Logistics-specific responsibilities are the following:

- Locate and provide requested supplies and equipment. Every community should establish a resource list of the special tools, equipment, and personnel available. This list should be carried in the command post or chief officers' vehicles. The lists should be shared by companies under the same mutual-aid agreements. As Command requests (or receives requests from Operations or Planning) for additional equipment, Command should inform Logistics and then let Logistics make the appropriate calls or take the steps necessary to procure the item(s). Logistics must do the following:

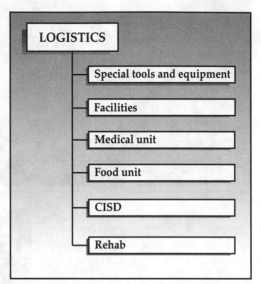

Figure 6-12. Sections under the direction of Logistics.

- —Ensure that the request is fully understood. If it is ambiguous or conflicting or if the request is for equipment that is unfamiliar, Logistics must take the steps necessary to understand exactly what is needed.
- —Check equipment lists for equipment not on the scene or within the department and make the appropriate contacts to obtain it.
- —Ensure that the equipment is en route and the provider knows where to deliver it. Calls may have to be made to the police, who may be in charge of access to the incident area, to allow the provider to enter the scene.[4]
- —Notify Command when the equipment arrives and lend any needed assistance to the provider. Normally, a firefighter is assigned to work with the outside provider.
- —Provide support for on-scene units. At large, long-duration incidents, on-scene personnel and equipment need many supplies to keep them going. Among them are fuel for apparatus, mechanics to make emergency repairs (mechanics should report directly to Logistics), air for SCBA bottles, hand tools such as axes and pike poles, and extra hoselines and nozzles.

Figure 6-13. Rehab is established to meet the needs of the crews at a large incident. This responsibility falls under the logistics section.

—Critical incident stress defusings. This is another area of support for on-scene personnel. These informal meetings allow the personnel to discuss the incident and its ups and downs and are considered precursors to more formal debriefings. The defusing team and its coordination are handled by Logistics, who most probably will also have to provide an appropriate location for the defusing.

• Provide support for operating special equipment.
• Provide adequate communications. For most of the larger incidents that require operations, planning, and logistics officers, communications probably will not be a problem—at least to the point that a communications officer must be named. However, on some occasions, it may be necessary to assign an officer the task of acquiring additional or alternate methods of communicating on the scene or to other locations. One such incident springs to mind. It was a subway derailment and subsequent mass-casualty incident in Philadelphia several years ago. At this incident, communications between the upper (street) level and the subsurface area, where the victims were, were hampered due to the radio frequency and equipment in use at the time. An officer from the communications bureau was called to the scene, and hardwired radios were strung. Before using the hardwired radios, specific individuals spaced at intervals within and outside the subway area were used to relay messages. At high-rise fires, Logistics may be asked to activate in-house phone systems or even to string hardwired systems up towers or shafts. Logistics may be the one who would go around an incident collecting

every portable radio and unused cell phone on the scene for use elsewhere. As the need arises, ingenuity surfaces.

Medical Unit

Some sources say that rehab[5] and victim treatment in the form of field hospitals fall under Logistics. Other sources say only rehab falls under Logistics and all victim medical needs come under Operations. In my opinion, either is correct as long as the lines of responsibility do not get crossed. I believe rehab is best suited to Logistics. I prefer to have special medical units, such as mobile field hospitals or DMATs (disaster medical assistance teams), fall under Logistics. There are certainly logistical problems associated with acquiring, setting up, and operating these sophisticated units. All triage and nonlife-threatening treatment can come under the supervision of Operations. An acceptable rule of thumb is, If we carry it on our rigs and are trained to provide the specific service, it probably should fall under Operations. If the task is beyond the training of the department and the equipment or tools need to be brought to the scene, it should fall under Logistics. Having a specific plan and conducting occasional disaster drills will go a long way in helping to identify flaws and pitfalls in individual systems and determine the most efficacious reporting structure.

Food Unit

A food unit was established at many incidents to which we responded. It wasn't called a food unit many years ago; it was called the Red Cross, Salvation Army, or canteen. In the old days, a chief at the incident sometimes called for these organizations—other times, they just showed up. A good logistics officer can plan and call for the establishment of such units. After it has been determined that food should be brought to the scene (and after consultation with Command), Logistics can determine the appropriate location for setting up the unit. The time to stop the food unit should be determined by a joint discussion between Operations (who forecasts future staffing needs for the incident), Planning (who should know where the situation actually stands), Command (who should have the overall big picture), and Logistics. Again, if Command chooses not to designate Logistics (or any other of the four functions), then Command is responsible for these tasks.

A final word about Logistics: As stated above, logistics can become the "melting pot" of incident operations. If it doesn't fit, give it to Logistics. If a task or need arises and it isn't quite Operations (such as finding something to spread a neutralizing agent on top of a spill) and it isn't quite Planning (such as elevator use at a high-rise incident), give it to Logistics!

THE ROLE OF ADMINISTRATION

Of the four functions of Command, administration is best suited for Command, especially if Command is the chief of the department. Addition-

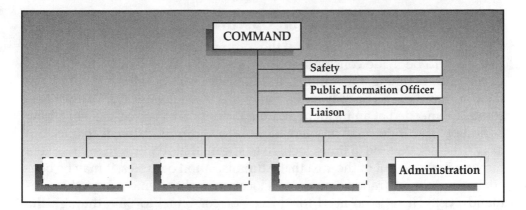

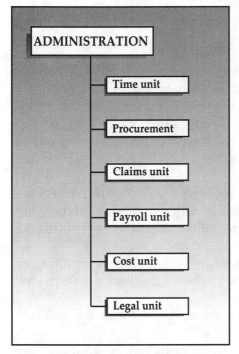

Figure 6-14 (above). Administration section. Figure 6-15 (left). Sections under the direction of Administration.

ally, it normally is the last function Command gives up. Administration represents the formal branch of the incident. Prior to the Consortium (as mentioned in Chapter 1), administration was called "finance." In Chapter 1, we learned that the first forms of the incident management system were devised by wildland firefighters. There are few organized wildland fire departments. At large wildland fires, professional wildland firefighters are brought in to work the fire. At "base," these firefighters check in and complete payroll records. That was the root of and foundation for finance. The Consortium recognized that it needed to readdress the role of finance to fit the more traditional forms of firefighting—hence, the name change.

Who Should Serve as Administration?

Most larger departments have a "numbers cruncher" in their administration. In the Toledo Fire Department, that position is referred to as the "administrative assistant." This individual has all the credentials for the responsibilities Administration needs to fulfill at a large incident. The administrative assistant should understand the department's financial picture and know how to puchase needed items. If departments do not have an officer responsible for day-to-day purchases, any chief or staff officer with a good knowledge of the jurisdiction's operations and good organizational skills will do.

Where Should Administration Operate?

Administration should operate at the command post. Most of the decisions that need to be made concerning purchases and claims require input from Command, Planning, Operations, and Logistics—all of whom should be at the command post. The only exception to this is when large numbers of personnel are needed at an incident and the incident is expected to span a long period of time. If the most problematic area for Administration is to track on-duty time or overtime (a time clock, so to speak) that requires many workers moving into and out of the area that Administration occupies, it may be beneficial for Administration to move to a location slightly remote from the command post. There is no need for dozens of personnel to pass through the command post and distract the "think tank."

Administration's Responsibilities

- *Time unit.* In an incident that calls for an operation of long duration, whether the department be volunteer or paid, an accounting of who is working and for how long is required.
- *Procurement.* This unit is responsible for keeping track of what is purchased and from whom. If deliveries of items are made to the scene of the incident, any documentation or receipt for such items should be sent to the administration officer for filing.
- *Claims unit.* This unit tracks personnel injuries. At a natural disaster, a location for a claims representative may be set up for civilians who want to report damage or need. The documentation for cost recovery will take place here.
- *Payroll unit.* If the incident is to be extended and a process for hiring or otherwise compensating workers needs to be established, then a payroll unit may be required.
- *Cost unit.* This unit is responsible for accounting for the actual and estimated costs of mitigating the incident.
- *Legal unit.* This is the newest aspect of the IMS and administration. I am not too sure that our interests are best served by having an attorney at the command post. Anyone who has been at the command post at a major incident knows that some of the conversations held there certainly might raise the concern of a lawyer representing the jurisdiction. However, for certain incidents, such as a haz-mat incident with technical legal ramifications, an attorney may prove useful. Unless your department's procedures stipulate that a representative from your legal department be present, you will have to request that an attorney be present.

Endnotes

1. Individual departments should establish an accountability system for all structure fires in accordance with NFPA 1500 standards. I will not elaborate on any such system. Many prototypes are available; to suggest a specific type would not be prudent.

2. Departments should consider requiring periodic PARs at large incidents. These would come at 15-, 20-, or 30-minute intervals.

3. Benchmarks are formal announcements made by sector officers. They are discussed at length in Chapter 19.

4. With the advent of cellular phones, there is no excuse for not having at least one telephone on the incident scene. Additionally, all chiefs' cars should have a current phone directory as well as an updated resource list.

5. Rehab (rehabilitation) is a sector assigned to provide on-scene members with an area to rest, replenish fluids, and have their basic and minor medical needs met. Blood pressure, pulse, and respiration should be checked and baselines established.

REVIEW QUESTIONS

1. List the four functions of Command.

2. These functions must be fulfilled at _____.

3. What is the role of the operations officer?

4. What are Operations' three responsibilities at an incident?

5. Where should Operations operate?

6. Why should Command be on a channel separate from Operations?

7. What is Planning's role?

8. What do you consider to be Planning's four most important responsibilities?

9. Where should Planning operate?

10. What is the role of Logistics?

11. Where should Logistics operate?

12. List two of Logistic's responsibilities.

13. What is Administration's role?

14. Where should Administration operate?

15. List two of Administration's responsibilities.

DISCUSSION QUESTIONS

1. Discuss or explain incidents to which you responded where Command delegated one or all of the functions. In your opinion, did this prove successful for the incident?

2. Assume you are the chief of a small department. List all of the functions of Command and indicate those you would delegate to mutual-aid chiefs without hesitation and those you would hesitate to relinquish. Give reasons and examples to support your answers.

3. Compare and contrast the roles of Operations and Command as they pertain to determining strategies at fires.

Section II

Staging and Sectorization

7

Staging

Section One covered Command: its identity, purpose, staff, and functions. Section Two will focus on the last two components of the incident management system (IMS): staging and sectorization. In my opinion, the command component is somewhat formal ("stuffy") and conceptual.

It certainly may be difficult for some to understand the concept of adminstration at a working fire. For firefighters and officers, staging and sectorization are more tangible, interesting, and identifiable concepts.

KEY CONCEPTS

✔ Assigned freelancing

✔ Level I staging

✔ Level II staging

✔ Staging

✔ Staging officer

✔ Still-box response

STAGING

Staging is the placement of uncommitted apparatus and personnel at an incident. Staging (along with command modes) is the emergency responders' playbook. Were it not for staging, responding units would have no specific focus prior to assignment. According to the National Fire Academy, staging areas, regardless of their type or size, must be located where "resources are ready for immediate assignment." Staging presents three positive qualities at every incident:

- *It virtually stops the initial freelancing of incoming units.* As units arrive at an

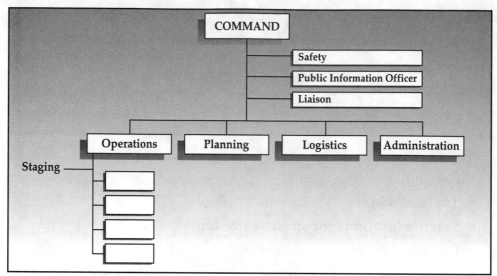

Figure 7-1. The chain of command for staging.

incident, they are required to stage, stopping them from going up to the incident and doing whatever they feel needs to be done. Staging will not stop assigned freelancing (assigned companies doing tasks other than those assigned to them). Only training will stop assigned freelancing. But staging should keep responding, unassigned units and personnel from approaching the scene until directed to do so by Command or Operations. Staging gives control to Command (or Operations, if designated) by allowing him to place incoming or staged units where necessary.

- *Staging sets a "calming" tone for the incident.* No matter what the size or scope of the incident, if staging is used, members are required to stop and await direction. In many ways, this slows incoming officers down and allows them to concentrate on their crews and where to stage and not on the incident or possible assignments. Once staged, the unit awaits direction from Command or Operations. There is no pressure to create your own assignment.

- *Staging allows one to truly "look" important.* No more rushing up to the scene of a false alarm on a narrow residential street and then fighting over who should back out instead of pulling ahead. Which of the following scenarios presents a more professional appearance to civilians?

SCENARIO 1

You are dispatched to a structure fire in a residential area in your community. The assignment is two engines, a truck company, a rescue squad, and a chief. The first unit arrives and reports "nothing showing." As other units arrive on the scene, they pull up as close to the address as possible (which, by the way, is in the middle of the block). By the time the first-in officer reports that this was a false alarm and that "everyone can go in service" (back to available status), about 10 firefighters are on the front porch of the house. At that exact time, Dispatch asks the squad (the second unit on the scene now blocked in by a ladder and an engine going north and an engine and the chief going south) if it can take another fire. If your community has squads, you know what will happen next. Or, what if there are no immediate runs, but everyone is ready to leave the scene? (Two weeks ago, after a rash of backing accidents, the chief of the department issued an order that any member involved in a backing-up accident will be charged.) No one wants to back up, and we end up arguing in the street. How professional do we look to the public in any of the above situations?

SCENARIO 2

What if, instead of all responding units running up to the scene, everyone responding except the first on-scene unit stays back at the nearest intersection and awaits for direction? How would that look to the public—like a group of anxious rookies, running around "trying" to look like they can save the world (even when there is absolutely nothing to save) or like professionals who are ready if needed? As stated above, staging creates the fire department playbook. In the past, members did what they thought was necessary at incidents. Things got done, and fires went out. (The fire always eventually goes out.) But think of how a football team would look or produce if there were no official plays and each member of the team did whatever he thought was important or necessary.

Types of Staging

There are two types of staging, depending on the level of response.

Level I Staging

Used at responses up to a first- or regular-alarm assignment.[1] In these instances, the first unit to arrive on the scene gives a report that includes the following information: unit designation, address, a brief conditions report, and the command mode.

Engine 5 is at 1945 Vermont. We have smoke showing. Engine 5 is Vermont Command.

On receipt of this announcement, all responding units should stage at an appropriate location and await assignment from Command.

In Level I staging, the officer of the responding unit selects the appropriate

Figure 7-2. An engine staged at a regular-alarm incident. The officer determines the location. Note the proximity to the hydrant. (Photo by author.)

location to stage. Appropriate locations for responding units in the nothing-showing mode would be as follows:

- *Engine companies:* The next-in engine should stage at or near a hydrant. All other responding engines should stage at an intersection in their direction of travel, unless that direction has been covered by another engine. In this case, it may be more prudent to go around the block in case the engine must come in from the other direction. In this case, Command should be informed that the engine is going around the block and staging at the (names of the streets) corner. The only exception to this is if the first unit on the scene is not an engine. In this instance, the first engine on the scene should go up to the fire address and stage in front of the building. This is done to get a quick water supply should a small fire erupt. However, when an engine company is required to pull up to the incident and stage because the first unit on the scene was not an engine, the crew should stage. Members should sit in the rig and await direction from the officer inside (Command). There's no need for three or four more firefighters to be walking through the house. If needed, Command will call for them.

- *Truck companies*: Responding truck companies should position themselves at an intersection that gives them the best access to the front of the building. If this means driving around the block, so be it. We have all complained when truckies stripped our rigs of tools because the truck had to park a block down the street because engines and squads blocked the front of the building. Truckies get upset when they have to lug their ground ladders, PPV fans, and other equipment a block because they can't get close to the fire—so give them access to the front of the building.

- *Squads:* Heavy squads or other rescue vehicles (units) can stage at the nearest intersection. If called up by Command, they can pull in a driveway or even off the street. The idea is to get out of the way and not come up until called.
- *Chief officers*: They can do what they want. It's that simple. When we first began using incident management, I used to go right up to the scene. Sometimes, I'd even get out of my car to see what was going on. (Remember, we are discussing the nothing-showing mode.) Now that I'm older and have more faith in my officers, I tend to stay back at the nearest intersection and wait to see if they have anything or if I'm going to get canceled.

In the fast-attack mode, the first-in unit participates in the incident. The next-in unit comes up and formally takes command. Once this happens, all other responding units should stage, according to the above guidelines. Remember, in the fast-attack mode, the first-in officer feels that his two hands will make a difference in the outcome of the incident. The officer will be inside participating with his crew.

In the command mode, once someone formally takes command, all units should stage as described above.

One frequently asked question is, What should happen on still-box responses such as alarm-system emergencies and garage fires where a response less than a regular alarm is dispatched? The same procedure holds true: The first unit should go up to the scene; the other responding unit should stage in a location advantageous for the type of situation and apparatus. If two engines are sent on an alarm-system incident (commercial building with a fire alarm sounding), then one unit should go to investigate the nature of the call and the other engine should stage in a location advantageous for supplementing the system if needed. If an engine and truck are sent on the same type of alarm and the first unit on the scene is the truck, then that crew should investigate and the engine company should stage at a location advantageous for supplementing the system. If, on the other hand, the engine arrives first, then the truck can stage in the front of the building.

This procedure applies to fire and EMS incidents. It works and must be enforced if the incident is truly to be controlled.

Level II Staging

Level II staging is used at incidents beyond the first- or regular-alarm response. The key difference between Level I and Level II staging is that in Level I, the officer determines the staging location. In Level II staging, Command or Operations chooses the site where incoming apparatus are to stage. In Level II staging, once the staging area site has been chosen, the responding units go to that area and not to the fire. In some instances, this can (and should) be blocks from the incident, and no smoke or fire may be seen from the staging area.

Once in the staging area, units are then sent to the fire as directed by Command or Operations. The staging area is under the direction of an officer, given the title of "Staging." All units in staging should report to the staging officer.

Chain of Command

The staging officer should report directly to Operations if an operations chief has been assigned. There is no need for Operations to have to tell Command to have Staging send an engine company to assist Division 2. That's a waste of time and breath for Operations and Command. Of course, if Command has not designated an operations officer, then Staging should report directly to Command.

Who Should Be the Staging Officer

The first unit to report to the staging area should assume the duties of staging officer. We have gone around and around on this. At first, we chose to have the first officer report to the staging area and automatically be designated as Staging. We have three-member truck companies. When a truck arrived first at the staging area, that meant that the truck officer would become the staging officer. That left a two-person truck company. It's hard enough doing truck work with three personnel, let alone two. We tried to get the truck crew another firefighter from the next unit in staging, but that didn't always work. We now have excluded truck company officers from taking on the duty of staging officer. Our procedures now state, "The first officer to report to the staging area (other than a truck officer) shall take the role of staging." If personnel on specific units in your jurisdiction prohibit you from proceduralizing that the first unit to arrive in staging automatically assumes the position of staging officer, those particular unit types should be excluded. Remember that as little time as possible should be wasted in establishing a staging officer.

I like to think of the staging area as "my own little fire department." To chief officers of smaller departments, staging may establish "your own 'larger' fire department." Units dispatched to staging are not available for other response. (If concurrent incidents occur, Command will decide whether staged units should respond to another incident or remain staged.) Next, I like to think of the staging officer as the "dispatcher" for my little fire department. Command or Operations asks for units and, unless specified, Staging sends whom he chooses, just as a dispatcher would.

Duties of the Staging Officer

The staging officer has three basic responsibilities:
- *To control the activities in the staging area*. Staging should ensure accessibility to and egress from the staging area. Placement of apparatus and some sort of accountability system should be established to ensure that

(a) units sent to staging are logged in and out for future reference, and

(b) members sent to staging remain in staging.

Units should not remain in staging for long periods of time. They will be assigned as needed or sent back to the station after the situation de-escalates. Facilities such as food or portable restrooms probably will not be needed, but if for some reason they are needed, Logistics would arrange for them.

- *To send the appropriate units to the incident (or other assignment).* Command or Operations normally will not worry about individual or particular units in staging. Except for specialized units such as aerial platforms or foam units, Command or Operations will ask for an engine, a truck, and so forth. It will be up to Staging to choose the appropriate unit and send it up to the incident (this will be explained in detail later).

- *To keep Command or Operations informed.* I would hate to be the poor staging officer who had been asked by Operations or Command to send an engine and had to radio back, "Staging to Command: No engines are left in the staging area." Staging periodically should call Operations and report the status of the units available in staging.

Staging to Operations: Be advised you have two engines and a truck left in staging.

Now that Operations has been informed of the units left in the staging area, Operations can confer with Command and discuss the need for additional units. If additional units are needed, Command—and only Command—should request them.

Staging should be monitoring the fire radio and attempting to keep up with the incident. If he can anticipate potential requests for units and knows that there may be a shortfall of those particular units, then Staging should inform Operations of the status of those particular units.

Staging to Operations: Be advised that if more than two aerials will be needed, you have only two left in staging.

That notice should suffice and take the responsibility of notification off the shoulders of Staging and place it on Operations or Command.

HOW STAGING SHOULD WORK

Assume that an operations officer is designated at this incident. Pretend that you are Staging at a large warehouse complex fire and are the officer of an engine company. Your engine was the first unit to report to the staging area. You arrive at the designated staging area and inform Operations that you are at the staging area.

Engine 4 to Operations: We are at the staging area. Engine 4 officer will be Staging.
Operations to Engine 4: Okay.[2]

We now have a staging area with a staging officer and one staged engine (4). There's no sense in not using a perfectly good engine and the remainder of your crew.

The remainder of the second-alarm assignment is two more engines, two truck companies, a heavy rescue squad, and a battalion chief. The chief will respond and report directly to the command post, as procedure dictates, unless otherwise directed by Operations while en route.

Operations wants to set up a ladder tower in the rear of the building.

Operations to Staging: Send a ladder and engine to Sector 3 and have them set up a water tower. I want them to try to cut off the spread to the west wing of the building.
Staging to Operations: Okay.

Staging should now look at the units in staging (or still en route to staging) and consider the following:
- the request made by Operations,
- the units available (in staging or en route), and
- the anticipated future needs of the incident.

Operations wants an aerial to use as a ladder pipe. Nothing fancy—just a straight stick. As Table 7-1 indicates, you have an aerial platform that can do

Table 7-1. INITIAL UNITS DISPATCHED TO LEVEL II STAGING
STAGED UNITS

Units	Capabilities	Staffing
Engine 4	1,500 gpm	Three personnel
Engine 7	1,250 gpm	Four personnel
Engine 6	1,500-gpm squirt	Five personnel, including a two-person rescue unit (two-passenger van)
Truck 17	100-foot aerial tower with a 1,000-gpm nozzle	
Truck 4	85-foot straight stick aerial with prepiped 1,000-gpm nozzle	
Squad 1	Heavy rescue	Four-person crew
Battalion 3	One car	One chief

many things and safely carry several firefighters and a straight-stick aerial. Both can throw the same amount of water. Reach is no factor at this time. Choose the straight stick. Operations asked for an engine to supply the truck. All three engines staged can provide the capacity of the truck you have chosen plus water for a handline if needed. One has three personnel, one has four firefighters, and one has five firefighters. How many firefighters does it take to supply a water tower? No more than three, or your department needs a lot of training. I'd send the engine with three firefighters.

After Staging makes his decision, he should inform the units he will be sending and provide them with any additional necessary information. ("The chief wants you to set up in Sector 3 and try to keep the fire from getting into the west wing of the fire building.") Finally, Staging should tell Operations who was assigned.

Staging to Operations: You're getting Engine 4 and Truck 4 for Sector 3.

Now, Operations, Command, or their aide can update the command board, indicating that Engine 4 is feeding Truck 4 in Sector 3.

Table 7-2. UPDATED STAGED UNITS (AFTER OPERATION'S FIRST REQUEST)

UPDATED STAGED UNITS

Units	Capabilities	Staffing
Engine 7	1,250 gpm	Four personnel
Engine 6	1,500-gpm squirt	Five personnel, including a two-person rescue unit (two-passenger van)
Truck 17	100-foot aerial tower with a 1,000-gpm nozzle	
Squad 1	Heavy rescue	Four personnel

Next, Operations calls for an engine company to assist Engine 3 with attack in Sector A.

Operations to Staging: Send a company to assist with attack in Sector A. Engine 3 is Sector A attack.

Two engine companies, a truck, and a squad company are left in staging (see Table 7-2). Operations wants personnel, not equipment. I'd send the squad. Again, Staging should notify the squad officer of the assignment and tell him to whom he should report. ("The chief wants you to assist attack in Sector A. Captain Douglas from Engine 3 is the attack officer.")

Staging should tell Operations who is being sent. ("Staging to Operations: Squad 1 is being sent to assist attack in Sector A. Operations to Staging: Okay on Squad 1.")

Table 7-3. UPDATED STAGED UNITS AFTER OPERATION'S SECOND REQUEST

UPDATED STAGED UNITS

Units	Capabilities	Staffing
Engine 7	1,250 gpm	Four personnel
Engine 6	1,500-gpm squirt	Five personnel, including a two-person rescue unit (two-passenger van)
Truck 17	100-foot aerial tower with a 1,000-gpm nozzle	

At this point, Staging probably should tell Operations the status of staging.

Staging to Operations: We're down to two engines and a truck in staging. Operations to Staging: Okay.

Now it's up to Operations to look at the incident and determine if additional units may be needed. If so, he should tell Command what he will need to be staged.

WHAT THE STAGING OFFICER SHOULD TELL ASSIGNED UNITS

After Staging determines what unit(s) he will be sending to the scene, he should inform the officer or person in charge of the unit, preferably in a face-to-face conversation. Items covered in this brief exchange should include the following:
- what the assignment is (attack, for example),
- where the assignment is (Sector 3, Division 16, or both,[3] for example),
- to whom to report if someone else is in charge of that assignment (Engine 7 or Lieutenant Smith, for example), and
- special instructions or equipment to take ("The chief wants you to let him know if the sprinklers are doing any good; make sure you check the plenum"; or "take a hose pack and pike poles with you.")

TOOLS NEEDED BY STAGING

Staging doesn't need a whole lot to do his task. A few necessities and a few "luxuries" will make life a lot easier, however.

Necessities
- Radio: mobile or portable (portable is better). Staging can move about the staging area and talk with staged officers.
- Something to write on and with: At times, requests from the command post may be quite explicit; it's easier to note who is (and was) in staging than to have to look around every time a unit or update is requested.

Luxuries
- A place to sit. If Staging's vehicle or an extra chief's car can be kept in staging, it would be advantageous to Staging. If a large incident occurs in bad weather, having a warm (or cool) dry place to operate from will be greatly appreciated.
- Lighting: At night, light makes it easier to read notes.
- An aide: Someone to help track staged units, act as a runner, listen to the radio, or do all three tasks sure would be nice.

A FINAL NOTE

When the command post determines that the need for staging no longer exists, a broadcast to all units and the dispatcher that "Staging is terminated" would be useful. The announcement should be logged by Dispatch or the sit-stat officer (planning) and on the incident time line.

Endnotes

1. In Toledo, a regular alarm consists of three engines, one truck, one rescue squad, and a battalion chief. Some cities have a different name for a similar response or add or lose a specific piece of equipment.

2. There is some question as to whether orders like this need to be repeated for the sake of clarification. In my opinion, that depends on the nature of the message and the amount of radio traffic at the time the message is given. If the message concerns life-threatening information or is very specific in nature and there is a lot of additional important radio traffic at the time, it probably should be repeated. One suggestion is that perhaps "understood" may be a better response than the typical "Okay."

3. The use of sector assignments is explained in detail in the next chapter. For now, don't worry about what they specifically mean.

REVIEW QUESTIONS

1. Define staging.

2. List the three positive qualities of staging.

3. Staging creates the _____.

4. What are the two types of staging?

5. On what factor does the type of staging depend?

6. At what level of response is Level I staging used?

7. Who determines where units will stage when using Level I staging?

8. At Level I staging, where should an engine stage?

9. What would be the only exception to the above?

10. Where should a truck stage during Level I staging?

11. Where should a squad stage during Level I staging?

12. Where should chief officers stage during Level I staging?

13. Does the location of staging change with command modes? Yes ___ No __

14. If sent as the second engine on an alarm-system incident, where should you stage?

15. When is Level II staging used?

16. Who determines the location of Level II staging?

17. Where do units report when using Level II staging?

18. What are the three duties of the staging officer?

19. What four things should the staging officer tell the units he is sending to fill Command's or Operation's request?

20. What things does Staging need at the staging area?

DISCUSSION QUESTIONS

1. Discuss how staging stops freelancing.

2. If your department uses staging, how do you choose the staging officer? If your department does not use staging, how could it best fit into your department?

3. Why can't the staging officer cut out the middleman (so to speak) and ask for additional units to be sent to staging?

8

Sectorization

Sectorization—*breaking down the incident into manageable units*—helps to define expectations and areas of responsibility. It is very closely related to three areas.

KEY CONCEPTS

✓ The "Big Four"
✓ Benchmarks
✓ Combination sector
✓ Functional sector
✓ Geographical sector
✓ Rakes
✓ Sectorization
✓ Sectors
✓ Soffit
✓ Tunnel vision

SPAN OF CONTROL

The effective span of control for emergency operations is considered to be three to seven, the optimum being five.[1] This is not a hard-and-fast rule. There will be times when sector officers will be responsible for more than five subordinates. I believe that if the responsibilities and roles of subordinates are not of a life-threatening nature or vital to the outcome of the incident, then an officer occasionally may have more than five subordinates.

CHAIN OF COMMAND

Sectorization defines in large part the chain of command of the incident (not necessarily that of the department). Lines of authority and responsibility for staff and sectors are defined. As the flowcharts indicate, the fire suppression sector officers operating under Operations cannot communicate directly with Command.

Conversely, Command should not talk directly with sector officers if an Operations officer is assigned.

TUNNEL VISION

The object of the IMS is to give everyone needed to handle an incident a specific assignment. Once that assignment has been received, the sector officer's job is to "tunnel in," or focus, on that task. If the sector officer is assigned backup, I, as Command, want that officer to concentrate on protecting the interior crews and any savable victims. I don't want that officer attacking the fire, venting the structure, or looking for fire to extinguish. That officer's responsibility is just to provide backup for interior crews. Sectorization ensures this focus by defining responsibilities.

TYPES OF SECTORS

Three types of sectors are used at all incidents, regardless of the incident type. The type of sector used depends on the following factors:
- whether the assignment is activity- or area-based, or
- whether the assignment is both activity- and area-based.

The three types of sectors are functional, geographical, and combination.

Functional

These sectors, which almost entirely fit under the operations officer, are characterized by a specific activity Operations wants performed. They are the "verbs" of the fire service—the action words such as attack, search, backup, extricate, triage, and decon. (Note that the list spans several types of incidents. The IMS fits all kinds of incidents.)

Geographical

These sectors, which define area, also fit almost exclusively under operations. They can represent area by location: inside or outside, the level in a structure, a vehicle type or location, and so on.

Combination

These sectors, under operations, are functional and geographic. Combination sectors make it possible to fine-tune tasks, specifically to define a task that should be conducted in a specific area, such as Division 2 attack or extrication in the green Pinto.

If Command assigns a company to only a functional *or* geographical sector, that company is responsible for all the activities necessary in the entire area or a specific task in the entire structure. For example, if Command assigns Engine 5 to Division 3, Engine 5 is responsible for *all* activities on Division 3. That may include attack, backup, salvage, and overhaul. However, if Command assigns Engine 5 attack, then Engine 5 is responsible for putting out all the fire on all the floors of the structure.

NAMING SECTORS

First of all, what's in a name? Absolutely nothing! All of the terms presented in this chapter are taken from other working forms of the IMS. One of the big problems with the IMS was that larger departments seemed to be "shoving" their terminology down the throats of others, sometimes smaller departments. This goes back to the discussion in Chapter 1. What are presented in the following paragraphs are suggestions and only suggestions. These examples are taken from Toledo, Phoenix, Los Angeles, and other cities that use the IMS. At least one example of what "something" can be called is given; in some cases, several examples are given. What a thing is called is not as important as the fact that everyone who could be responding to a particular incident knows what the term means. If you call the floor of a structure a floor, division, sector, or "rebasaurus," it doesn't matter. What is important is that the next-in unit understands what Command wants when he says report to such and such an area.

FUNCTIONAL SECTORS IN DETAIL

Functional sectors sometimes are referred to as "groups" (attack group, search group, and so on.) They are given specific tasks to accomplish. If Command or Operations simply designates a functional sector to a crew, then that group is responsible for accomplishing that task in the entire structure, if a structure is involved. As an example, if Command designates Engine 7 as attack, the officer of Engine 7 is responsible for putting out all the fire in the structure. All the fire! That includes checking for extension and overhaul. Engine 7—no one else—has the hose. Now, if Command designates that Engine 7 is attack and the officer of Engine 7 gets inside and determines that he can't handle all the fire in the structure (from his vantage point), he should get on the radio to Command and tell him so.

Attack to Command: I think we'll need another attack company on Division 2.

Now it's up to Command to decide if he wants to assign another company to help attack the fire. Command could make one of two basic transmissions.

Radio Command to Attack: I'll get another attack company on Division 2. You will now be Division 1 attack ...
OR
Command to Attack: It looks like you got it from out here! I'm low on companies out here. Take your time and work your way up to Division 2. If things change, let me know!"

With either of these two transmissions, Command's expectations should be clear to Attack.

GEOGRAPHICAL SECTORS IN DETAIL

Geographical sectors, as you probably suspected, divide the scene into specific areas. At fires at which it would be confusing to give direction to two companies fighting fires on different floors, this approach helps to differentiate between companies. Geographical sectors normally are referred to as "sector," "division," or "side," for example.

A question often asked is, Why not just call the company by unit number? The answer is simple. It is easier for Command (or Operations) to know what the assignments are than who is filling them. Besides, using sector assignments as opposed to unit designations reinforces expectations. (Example: "You're Division 1 Attack!").

Take this a step further. Suppose that this fire goes to a second and then third alarm and Command (or Operations) has assigned seven companies to attack the fire in specific areas of the building. Now, would it be easier to refer to these companies by their work locations or unit numbers?

SCENARIO 1

A company is calling Operations with vital information. Let's say Command has seven companies assigned to attack in seven areas of a structure. Operations then hears this transmission: "Engine 7 to Operations: We're trapped and out of water and air. Send us some help!"

The first thing Operations has to do is to remember where he assigned Engine 7. Is Command going to go over to Engine 7 and start to follow the line into the structure? What if Engine 7 didn't take its line off the engine? It may be as simple as looking at the command board, but it is an unnecessary step.

Now consider the same transmission with the use of geographical sectors.

Division 4 Attack to Operations: We're trapped and out of water and air; send us some help!

How much simpler it is. Command gets crews started to the fourth floor.

Which company is assigned to a sector becomes insignificant. What is significant is that a company is assigned to handle that particular aspect of the incident and that the officer and Command (or Operations) know which unit is involved.[2]

Geographical Sector Assignments

The following are offered as suggestions. When more than one example for the same area is offered, they are presented as options. The following examples already are used by the NFA, *FIRESCOPE,* the Consortium, *Fireground Command (FGC)* (Phoenix), and Toledo. I believe that the fire service as a whole should strive to adopt a single national IMS.

But until then, there are no hard-and-fast rules. The only rule is that everyone who is or could be responding knows what Command means when giving assignments.

Where to Start

As a rule, geographical sector assignments should be in relation to the command post. The command post (usually the area in front of the structure) is referred to as "Sector (or Side) 1." (An easy way to remember this is with the phrase "I'm #1.") The progressions then go in a clockwise rotation around the structure. Some departments always use the street address side of the building as the command post focal point. I have a slight problem with that. What if the command post is not located in front of the building? Additionally, some buildings in Toledo have more than one address on more than one street. If a fire occurs at one of those structures, where's the front of the building, and where is the focal point for all subsequent sector assignments? Make any geographical assignment in relation to the command post; if the command post is not going to be located in front of the structure, then Command is required to periodically announce the location of the command post.

Figure 8-1. Interior sector assignments are designated by letters.

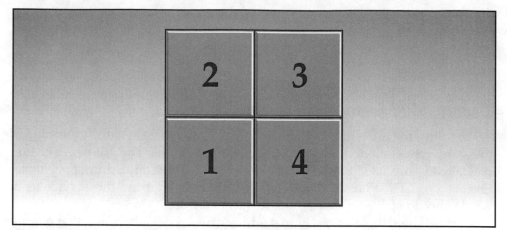

Figure 8-2. Numbers are used to represent interior sector areas.

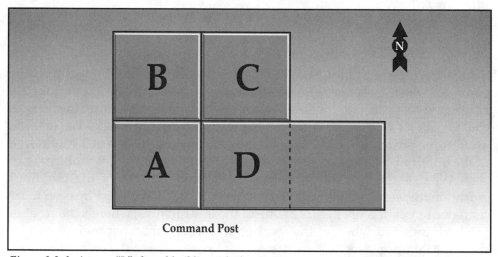

Command Post

Figure 8-3 depicts an "L"-shaped building. The key to structures such as this is that everyone knows what specific areas are called. We responded to a fire in an "L"-shaped structure that had one section one story high and another three stories high. We wound up referring to the different leveled buildings as separate structures. One was the "three-story building"; the other was the "one-story building." We had "three-story Division 2, Sector A attack" and "one-story Sector 2 extension" (checking for extension on the wall on Side 2 of the structure). This may sound confusing, but the confusion vanished as soon as crew members saw the building.

Interior of Structures

Figure 8-1 is probably the method most often used to divide an interior floor area. All sector designations are made in relation to the command post. In this example, the command post is in front of the building. Letters are used in this example. Numbers could also be used, as in Figure 8-2, but I believe numbers can be confused with floor levels in the building. Of course, there are few truly square buildings. Adaptations may need to be made.

Geographical sectors allow Command to fine-tune an assignment. Let's say the fire occurs in a very cluttered house. There are several fires (incendiary cause) and much overhaul. Instead of assigning one officer to the entire structure and missing some areas or allowing for flare-ups while working in other areas, Command can assign four companies to overhaul; each will be responsible for a specific area of the first floor. For example: Engine 5 has Sector A overhaul; Engine 13, Sector B overhaul; Engine 6, Sector C overhaul; Truck 13, Sector D overhaul. Marking boundaries of responsibility with scene tape is not a common practice. We are professionals and should avoid senseless squabbles. When each unit completes the task in its assigned area, it reports to Command by using benchmarks[3] and letting Command determine what the unit's next assignment should be.

Interior sectors can also be used for interior exposures, using the sectors to define areas to which the fire is traveling.

If necessary, each of these areas could be broken down into subsections, as indicated in Figure 8-4, an approach that would be very handy in a structure with a large floor area or complicated tasks.

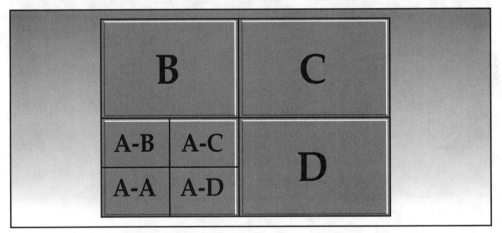

Figure 8-4. Example of interior sectors with subsectors in Sector A.

TERMS FOR OUTSIDE OF THE STRUCTURE (EXPOSURES)

I have heard exposure assignments defined as the following: "tan house," "house on the right," and "north exposure." The problem here is that they were all used within the same department. Terminology within a department should be standardized. Exposures can be designated by number, letter, or (my least favorite) direction (i.e., north, south, east, west).

Whatever the designation, it must be consistent and the opposite of the interior sector type. If you use letters (A, B, C, D, and so on) for interior sectors, use numbers (1, 2, 3, 4, and so on) for exterior exposures. If you use directions for exterior exposures, you can use numbers or letters for interior sectors. Again, whatever you use, be consistent; make sure everyone you respond with (including, and probably most importantly, mutual-aid companies) understands your terminology.

Remember: Sectors are assigned in relation to the command post. For most fires, we usually operate in front of the house. At commercial occupancies, this may not hold true. Weather and other conditions, such as wind or heavy smoke, may also alter this. In this form of designation, the command post is in front of the structure and is always #1. With the command post as #1, revolving clockwise, the exposure on the left is "Exposure #2." When Command wants to assign a company an exposure (the garage in the rear, for example), he doesn't have to be concerned with direction, color, or whether it's to the right or left. The rear is always "3" (bearing in mind the command post). Another way to put it is, "Exposure 3" is always opposite the command post. Incoming officers, on hearing Command assign them Exposure 3, should automatically visualize the rear area of the structure. What's easier, "Engine 7, take the exposure in the rear of the building" or "Engine 7 is Exposure 3"?

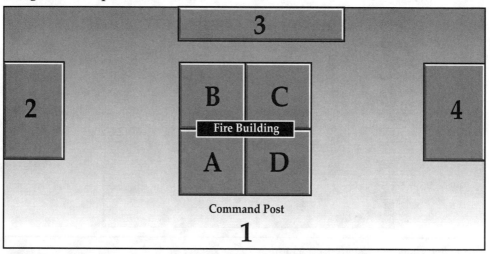

Figure 8-5. Exterior exposures designated by numbers.

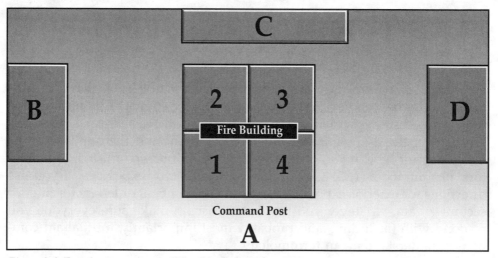

Figure 8-6. Exterior exposures designated by letters.

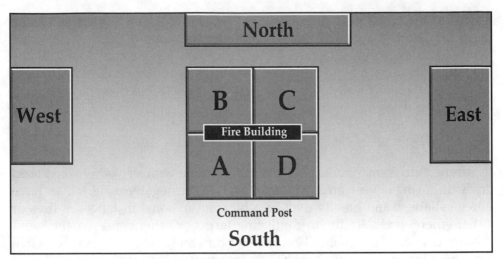

Figure 8-7. Exterior exposures designated by directions.

I personally do not like to use "direction" for several reasons. First, I am terrible at directions. My wife usually pulls us through on trips when I get us lost. I can't take her on every fire to tell me where "north" is. Second, there is only one true north-south street in Toledo—Canton Avenue—and it's only a few blocks long. Most of our streets run untrue to direction. I'd hate to have to assign a south-southwest exposure. This situation may not hold true in rural America. Many country roads are true north-south or east-west roads. Direction may work quite well in rural America. Finally, the city refuses to issue compasses to the fire department. Give me (and all other rigs) a dashboard compass, and maybe I'll consider using direction.

DIFFERENT FLOOR LEVELS IN A STRUCTURE

After selecting the methods for differentiating between interior and exterior areas, a method for distinguishing between floors in a structure must be chosen. In my experience, differentiating between floor levels is beneficial even at fires in single-family dwellings. We customarily search all floors of a structure, including the basement and attic (if the attic is accessed by stairs). To assign searches by probability (priority), it may be necessary to start a crew on one floor and the next available crew on another floor. This differentiation between floors defines responsibility and helps Command with the prioritization process.

Command to Squad 7: You're Division 3 Search.

The terms most commonly used to differentiate between floors are division, sector, and floor. Chief William Goldfeder of the Mason-Deerfield (OH) Joint Fire District has suggested to me that some East Coast fire departments (such as the City of New York {NY} Fire Department) may hesitate to use the term "division" to refer to a sector. Goldfeder spent many years of his career working in departments in the New York/New Jersey area. Many of the departments use the word "division" as a rank within their departments, as in "division chief." Using "division" to refer to floors in these departments may be confusing.

Division

For me, "division" works best. It doesn't rhyme with any other word used to distinguish between areas in a structure. On the negative side, it does have three syllables, and brevity may be a blessing in some instances at fires or other emergencies. If, due to conflicting terms or other concerns, the word "floor" is your department's best term for differentiating between levels in a structure, insert the word "number" before or after the number or level—for example, "number 3 floor," "number 4 floor," or "floor number 3," "floor number 4." This seems to stop some of the confusion with rhyming words.

Sector

The problem I have with using this word to distinguish between floors is that it seems to be used to death in the incident management arena. On the positive side, again, it does not sound like any other word that could be used for floor.

Floor

As far as I am concerned, this is probably the worst word that could be used. It is also the one that Toledo firefighters use the most. My main concern is that it rhymes with "four," as in "floor four." On the positive side, it is easy to remember.

So there you have it. As long as it works for you and everyone responding knows what you call a specific floor, any word will do.

EXTERIOR SIDES OF THE FIRE STRUCTURE

At times, Command or Operations may have to assign a company to work on the exterior of a structure, such as when exterior wall surfaces need to be overhauled or roofline areas, such as eave soffits or rakes, need to be overhauled or removed. Command should also refer to these areas by specific designations. Normally, they will be the same numbers or letters (or direction, if you must) as for exposures. The difference is that exposures are referred to as "Exposure 3," and exterior wall surfaces are referred to as "Side 3." If a soffit must be removed, a company is assigned to work on Side 3. Command can give a more specific location face to face or over the radio.

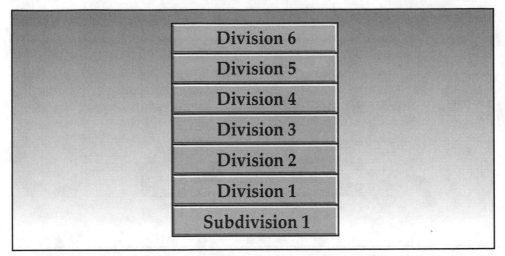

Figure 8-8. Floor designations using divisions.

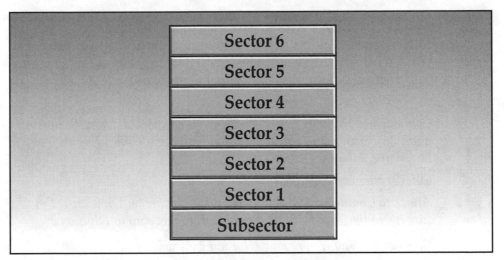

Figure 8-9. Floor designations using sectors.

 Engine 5: You're Side 3. Concentrate on the eave soffit where Sides 2 and 3 intersect.

Close your eyes. Try to picture what I just said. Now look at Figure 8-10, and see where Command would have wanted you to operate.

COMBINATION SECTORS

Sector assignments can be a combination of functional and geographic sectors. This allows for fine-tuning assignments. If Command determines that more than one attack line is needed to extinguish a fire, a method for distinguishing between these companies must exist. That's where combination

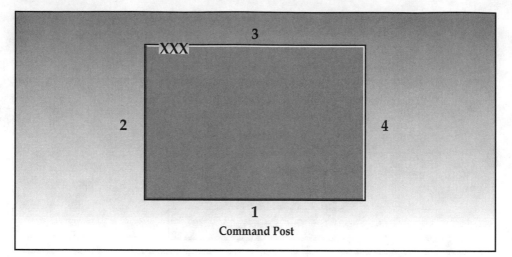

Figure 8-10. "XXX" indicates area at which Engine 5 was expected to operate (near the intersection of Sides 2 and 3).

sectors come in. "Division 1 Attack" specifies that Command has given the responsibility of fire attack on the first floor to a particular company. The only way Division 1 Attack can move itself and its line to attack fire on another floor in a structure is (1) at the direction of Command:

> *Command to Division 1 Attack: After you've knocked down the fire on Division 1, move your line to Division 3. I have assigned Engine 6 as Division 2 Attack.*
>
> *OR, (2) after confirmation from Command:*
>
> *Attack to Command: We've knocked down the fire on Division 1. I'm sure it's in the floor above us. Permission to move up to Division 2.*
>
> *Command: Okay. You're now Division 2 Attack.*

Table 8-1 indicates some more popular (frequently used) terms, where they usually fall, and their meaning.

Examples of combination sectors are Division 2, Sector A Attack; Division 14 Search; and Division 4, Sector A Subsector A Overhaul.

Note: If your department currently does not use any form of sectors for fires (or other emergencies), I am providing here what the National Fire Academy advocates (based on *FIRESCOPE).* In fact, the NFA discourages the use of the "S" (sector) word. I realize they do not specifically fit the way I, or we in Toledo, refer to sectors. I offer them as suggestions only. Unfortunately, there is no standard method used across the United States.

- All functional sectors are referred to as "groups," such as attack group, vent group, search group, and so on. Groups transcend interior or exte-

Table 8-1. COMMAND MODE, UNITS ON-SCENE, ANNOUNCEMENT

Name	Indication	Comments
Sector	Generic term sometimes applied to floor level or outside area. Also used with geographical sector, such as #3 sector.	Too generic. Has too many meanings. I suggest its use be avoided.
Division	Generally indicates floor level. The National Fire Academy uses "division" to indicate any geographical sector.	Can be confused with a level or rank within some departments. If used as a geographical sector alone (not a combination sector), that officer is responsible for all activities in that area.
Group	Generally refers to functional sectors, such as "attack group."	Comes from the NFA and *FIRESCOPE.*
Branch	Helps subdivide large operational areas, where span of control would be violated. With a ventilation branch, for example, the roof would be divided, and each area would be under a specific officer. All officers would report to the ventilation branch officer.	Extremely helpful at large incidents.
Task Force	Any combination (usually five or fewer) of single units (companies) put together for a temporary assignment. They operate under a single task force leader.	Used in wildland firefighting and at major incidents.
Strike Team	The same type of apparatus or unit acting as one unit—three engines assigned as strike team #3," for example. Can be more than five units.	Used in wildland firefighting.

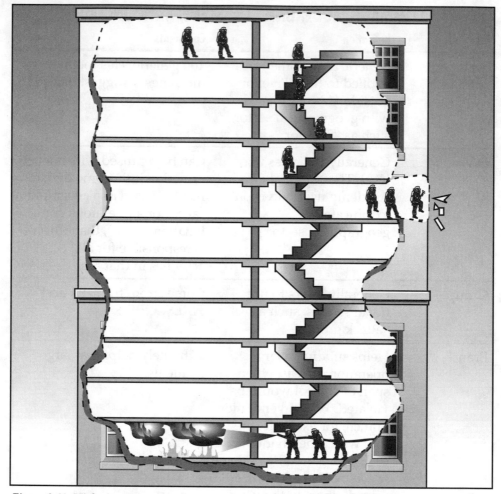

Figure 8-11. High-rise assignments.

rior areas. This means that a group will be responsible for its function in all areas of the structure or outside the incident area. If the span of control is going to be violated within a group, that functional sector is expanded and called a branch. If more than five units would be operating under the ventilation group (as with that large fire to which I responded mentioned in Chapter 1), they would be referred to as the ventilation branch.

- All geographical sectors are referred to as divisions. According to the NFA, "When a division is created, the division supervisor is responsible to consider all problems and solutions for that geographical area."[4] Normally, due to the nature of geographical sectors, no term is required to denote expanded operations within a specific division. If a crew is designated Division 3, that crew is responsible for anything needed on the third floor. (That includes attack, ventilation, or salvage that *has not been assigned as a group*.)

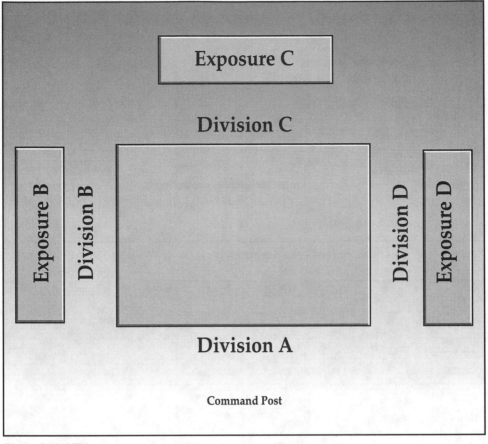

Figure 8-12. NFA sectors.

Note: According to the NFA, any group (or functional sector) assigned moves throughout the building (or area). Divisions take care of all else (other than the functional sectors already assigned) within their area.

Lastly, divisions and groups operate on the same level in the organization. Divisions do not have "authority" over groups, nor do groups have authority over divisions. They are equals.

REVOLVING SECTORS

A few years ago, we experienced several multiple-alarm fires. We also experienced something unwanted at these fires. It seemed that the initial officers at the fire established adequate strategies and began to sector off the fire in an acceptable manner. After a while into these fires, for one reason or another, Command (not the original Command) decided to move the location of the command post. As the command post was moved, so were the sector designations. What had been Sector A was changed to Sector B, but the name of the company in Sector A stayed the same. If you're confused reading this, imagine how the crews inside felt.

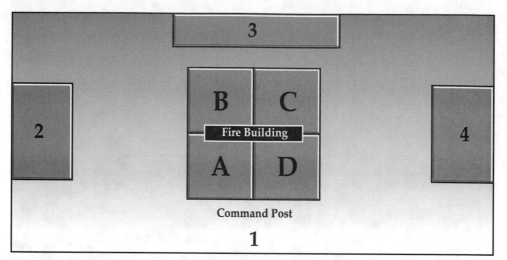

Figure 8-13. Example of the original sector designations.

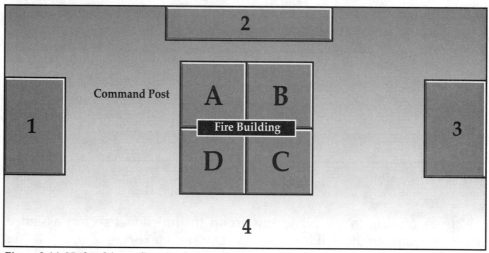

Figure 8-14. Updated (moved) sector designations.

A rule: The original designation of a geographical sector should not be changed during an incident.

Figure 8-13 shows the original sector assignments. A Sector A Attack and a Sector B attack were assigned when the decision was made to move the command post.

Figure 8-14 indicates the new sector designations. Imagine the confusion of the crews operating as Sector A Attack. When Command moved the command post, the sector designations changed. The company that was Sector A Attack suddenly was operating in the wrong area. Was the company expected to pick up its lines and move to the new area? Of course not! Our officers started calling this phenomenon "revolving sectors," and it was confusing and unnecessary.

Figure 8-15 notes the way it should have looked after the move of the com-

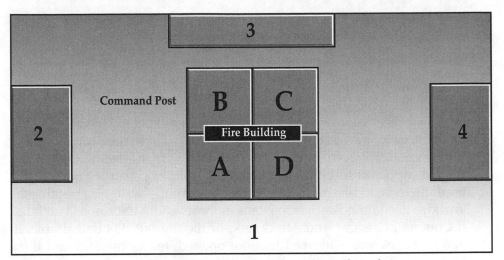

Figure 8-15. Sector designations after the command post location was changed.

mand post. Only the command post moved. The original sectors remained the same.

Changing fire or weather conditions sometimes may necessitate that the command post be moved; however, we've determined that it is much easier for Command or Operations to rotate the command sheets than it is for interior crews to pick up and move. There is no need to change anything but the angle of the command board.

RESPONSIBILITIES OF SECTOR OFFICERS

As with Command and his staff, sector officers have certain responsibilities. We hope our officers have been trained to fight fires as "we" would fight them: Give a good officer a task, define your expectations, and let him go for it! Sector officers have three basic responsibilities:

- *The safety of their crews.* First and foremost, sector officers should watch out for the safety of their crews. Chapter 5 addresses the role of the safety officer. The safety officer is indeed responsible for scene safety, but he can't be everywhere. To that end, the primary responsibilities of a sector officer are his safety and that of his crew.

If a crew member gets in trouble, we know where the rest of the crew will flock: toward that crew member. If that happens, where's the focus? Sector officers must resist the temptation to hold the nozzle or swing the ax. Their job is to direct and watch over. The officer holding a nozzle looking for fire to hit can't be concentrating on where the fire is, where the fire is going, where the fire came from, the atmosphere in the area, the next actions that need to be taken, the route for escape, whether he needs additional personnel, and so on. If the officer is holding the nozzle, he probably is tunneling in on only a specific area in a room, not the entire room or adjacent areas.

Finally, we are there to be part of the solution, not part of the problem. If a member gets hurt, we will focus on that firefighter until he is out of danger. During that time, we have let the owner of the structure down. We stopped focusing on the owner's property and instead focused on ourselves. If a sector officer can look out for the crew's safety and let the members do the actual work, no one should be let down—not the owner of the property, not Command, and not ourselves.

- *Accomplish the task assigned.* A sector officer's next concern is to accomplish the task he was sent to accomplish. As stated earlier, tunnel vision is now good. Think of it. If all sector officers are in, tunneling in on what they're supposed to be doing (i.e., attack, backup, search, and ventilation), there is no reason to worry about anything but their specific job. Truckies need no hose, Search needs no hose, and Attack needs no fans. If Command assigns you Attack, go put the fire out. It's that simple. If you're Attack and you need the roof opened, tell Command and then back down the stairs until you hear the axes or saws on the roof.

- *Keep Command informed.* Remember, Command should be at the command post. The days of a chief's sticking his unmasked head inside a second-floor bedroom to "see what it looks like" are over. Command needs to be updated on specific happenings during an incident. Some of those happenings are handled by benchmarks.

Information that must be relayed to Command includes unexpected findings such as multiple fires, the locating of victims, and the progress (or lack thereof) of ventilation.

In the National Fire Academy "Strategic Analysis of Fire Department Operations" course, wherein several simulated fires are fought, students are shown what it is like to run a fire at which gathering information is like pulling teeth. Selected sector officers are pulled aside prior to the simulation and told not to provide any information unless Command or Operations specifically asks for it.

The poor individual selected to be Command (or Operations) for that incident has a very hard time determining what is going on and the progress of the incident. Immediately after that scenario, the same student runs an incident in which all sector officers provide sufficient information throughout the incident. The difference between the two incidents is remarkable.

The rule is to keep Command informed about anything pertinent happening inside. As previously stated, benchmarks handle the majority of these happenings quite nicely. Command will let you know if you're talking too much.

THE "BIG FOUR" SECTORS

At most structure fires, some form of four functional sectors is almost always used. I refer to these functions as the "Big Four." In assigning these sectors, Command must look at the picture to develop his to-do list and then assign the Big Four as personnel become available. One or more of these functional sectors may not be required at some fires, such as a small incident

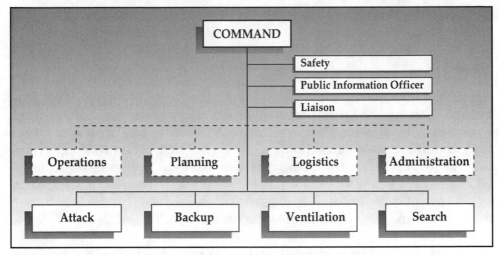

Figure 8-16. The "Big Four."

at which a single company can handle one or more of them. Conversely, more than these four sectors may be needed at some fires, and more than one company may be required to handle a specific sector.

Attack

No matter what the size of the fire, from a wastepaper basket fire to one that involves an entire floor, someone must put out the fire.

Search

We must always search for victims in structures—even vacant ones. We must not rely on the word of civilian bystanders or even occupants. Even if they report that everyone is out, that is no guarantee. A potential rescuer, such as a civilian, policeman, or family member, may have slipped unnoticed into the structure and have become trapped, disoriented, or overcome. Additionally, the homeless problem has placed an even greater emphasis on the need for primary searches. Even in rural America, the number of homeless or transient individuals has increased. They seek shelter and protection in vacant or unoccupied buildings. Based on fire conditions on arrival, the occupancy type (vacant, warehouse, barn, etc.), and the information given by occupants and neighbors, we must place search somewhere on our to-do list. It may not (and in many instances need not) be at the top of the list, but a search needs to be done in every structure in which we fight a fire.

Backup

NFPA 1500 and common sense dictate that we look out for and protect working crews operating inside at structure fires. Backup is a relatively new concept, especially at the normal residential structure fire. However, a backup sector should never be overlooked. Backup is explained in detail in Chapter 13.

Ventilation

If there is a fire, then in some way we must remove the by-products of combustion. This can be accomplished with an operation as basic as opening windows or by stacking PPV fans and selectively venting areas of a structure. The situation will and should dictate when ventilation will pop up on Command's to-do list. The question is, Do I have the personnel on-scene to ventilate when I would like to ventilate? Ventilation begun too early (except in the case of a high-heat, low-oxygen smoldering stage or backdraft fire) may spread the fire. Ventilation started too late in the incident may be a waste of time.

All of the above functions must be done in some fashion. At a rubbish or stove fire, the attack sector normally can put out the fire, open some windows, and look around to ensure that all occupants have left the area while still ensuring their own safety. As a matter of fact, this scenario (or a fashion of it) is the norm at most of our fires. Practice makes perfect. If we learn to delegate and divide responsibility at the "normal" fires, then on those rare occurences (large incidents) when responsibility must be delegated and divided, the task will be easier.

Do not simply assign the "Big Four" sectors automatically without regard to the situation or availability of on-scene units. It is ineffective and can be counter-productive. I have seen officers pull up at a structure fire and make four quick sector assignments to units that may not even have reached the scene. That's not commanding an incident. That's not an effective use of personnel or even of the gifts that God gave us. We could get robots to do that.

An effective IC will look at the picture and then make logical assignments, in order, as crews report on the scene. What was at the top of the to-do list a minute ago may be last on the to-do list a minute later because some vital piece of information has been gained. Continually changing assignments of responding or on-scene units signifies confusion and does not paint a picture of confidence of whoever is running the show. You have to look—and sound—important to be important! Under normal circumstances, it is coun-terproductive to assign a sector to a company that isn't on the scene yet. I learned this the hard way. At one incident, I ordered an en route truck to ven-tilate with the PPV fan. By the time the truck arrived on the scene, I needed its crew on the roof.

A final word on sectors: In preparing for promotional exams throughout my career, I have read many textbooks. One thing many of them had in common was a litany of the responsibilities of engine companies, truck companies, and squad companies. There was some crossover in these lists. Some lists were identical; others were varied and quite different. I understand that engines have a lot of hose and water, with the capability of acquiring and supplying even more water. I am equally aware that trucks have a lot of ladders (some portable and others more permanent) and a whole bunch of tools. What always confused me was the sense of "finality" of the purpose or responsibility of those specific units. What if a truck arrives first on the scene of a fire that has people trapped? Do crew members not go in because they don't have a line? Should they put a hole in the roof because "it's on their list"? What if a fire occurs and no truck is available? They're all out of service at another incident or responding from another community 15 minutes away and the engine crew can't make the stairs due to the heat. Do we not vent?

The thing I like about the IMS is that assignments are made according to Command's to-do list and, for the most part, not the apparatus type. If ventilation pops to the top of the list and no truckies are available to put on the roof to vent the house but an engine company is staged, put the engine crew on the roof! All engine specifications should stipulate ground ladders. A 28-footer should reach most two-story rooflines.

If people are trapped and only a truck and an engine are on the scene, Command should assign the engine as backup and the truck to search. Command should wait for more help to arrive to assign the things that need to be done later.

This is a versatile system. The only "hold barred" is that no one may do anything until Command says to! Along with staging, which gets unassigned personnel out of the way, sectorization is a very useful officer's tool. It sets actions in motion and defines responsibility. It gives Command the ability to focus on the whole incident.

Endnotes

1. *Managing the Fire Services, Second Edition.* 1988. International City Managers Association, Washington, D.C.

2. Whenever an assignment is made, the officer assigned should repeat the assignment back to Command (or Operations) so both know the message was understood. (Example: "Command to Engine 5: You'll be Division 3 Attack." "Engine 5: Okay on Division 3 Attack.")

3. Benchmarks are announcements that a particular activity has been completed. They are discussed in detail in Chapter 19.

4. "Command and Control of Fire Department Operations at Multi-Alarm Incidents," Instructor's Manual and Student Guide. National Fire Academy, Emmitsburg, Maryland. August 1995.

REVIEW QUESTIONS

1. Define sectorization.

2. Sectorization is very closely related to what three areas?

3. What factors determine which type of sector is used?

4. What are the three types of sectors?

5. What is the characteristic of functional sectors?

6. Give three examples of functional sectors.

7. What do geographical sectors define?

8. Combination sectors can be _____ and _____.

9. A designated functional sector is responsible for accomplishing the task _____.

10. All sector designations are made in relation to _____.

11. Indicate interior sectors by letter in the diagram.

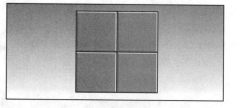

12. Indicate each interior sector by number in the diagram.

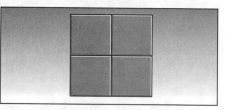

13. Indicate by letter interior sectors and sector and subsector assignments.

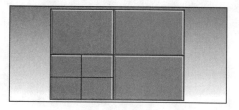

14. Indicate by number exterior exposures in the diagram.

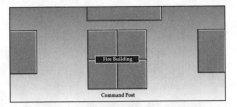

15. Indicate by letter exterior exposures in the diagram.

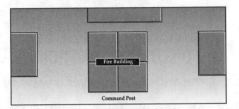

16. Indicate exterior exposures using direction.

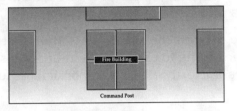

17. In the diagram, differentiate between floors using divisions. The lowest level represents the basement.

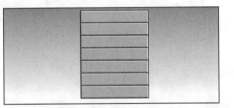

18. In the diagram, differentiate between floors using sectors. The lowest level represents the basement.

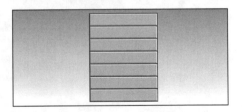

19. What is another name for functional sectors?

20. What is another name for the four functions of command and staff positions?

21. Define a strike team.

22. Define a task force.

23. Define a division.

24. What are the three responsibilities of sector officers?

25. List the "Big Four."

DISCUSSION QUESTIONS

1. Why do you think the author cited several ways to sector a structure?

2. Divide the class into groups (or individuals) and determine what method of sectoring a structure (interior and exterior, including exposures) best fits your department.

3. State a case to support the author's assertion that it is better to make assignments according to Command's to-do list instead of according to apparatus type or a textbook list of responsibilities for different company types.

Section III

Mission Statements

9

Defining Mission Statements

TOPICS

* **Mission statement**

KEY CONCEPTS

✓ Mission Statement

As stated in the first section of this text, the Toledo (OH) Department of Fire and Rescue Operations began using the incident management system (IMS) in 1988. We struggled through its implementation for several years. Something was still wrong. For as long as I can remember, prior to the IMS, officers did what they *thought* needed to be done at a fire or other emergency unless otherwise directed. An officer could pull an additional line; stretch it inside; and (a) attack the same fire another company was already fighting, (b) drop the line inside and begin a search, (c) drop the line and begin to ventilate, or (d) drop the line and look for friends and talk.

Another situation began to surface. We were getting complaints from crews that other companies were freelancing inside after having been given assignments. Companies assigned to backup were putting out fire. Companies assigned to search were opening windows. There was no control. To be blunt, officers weren't doing what was expected of them. The transition from freelance to total control wasn't all that smooth.

After much complaining and many meetings, we came up with the idea of creating mission statements that would define the purpose of the sector assignments. Mission statements define specific tasks a person or group of persons is to perform. I like to think of a mission statement as the reason for our existence. In the late 1980s, it was fashionable for fire departments to cre-

Figure 9-1. If no assignments have been made over the radio before you arrive at the scene and this is what you see when you walk up, who's doing what? What would the crew on the porch roof be doing?

ate and display their mission statements on letterheads, on business cards, and at the beginning of annual reports. Most included terms such as "to protect life and property" or "to provide fire protection consistently." Although at times wordy and a bit flowery, mission statements helped us to define our basic goals and the reasons we do what we do.

When we introduced the mission statements as they related to sector assignments, it was like a lightbulb had come on in our officers' heads. It became clear that when an assignment was made, it was the officers' task to complete that assignment and not to focus on other aspects of the incident. The next several chapters examine the mission statements of the most frequently assigned sectors.

REVIEW QUESTION

1. Define "mission statement."

DISCUSSION QUESTIONS

1. Compare the mission statements used by different fire departments represented by students in your class. If your department does not have a mission statement, create some examples for your department.

2. In groups or in individual essays, explain the significance of having separate mission statements for individual sector assignments.

3. Do you believe that individual mission statements could make a difference in your department's operations?

10

The Mission of Command

KEY CONCEPTS

✓ Exclusion zone

✓ Preplans

✓ Size-up

The mission of Command is to *coordinate* emergency crews' activities, making every effort to use accepted strategic concepts to effectively protect life and property from the effects of fire and other emergencies.

"Coordinate" is stressed because Command should not be a hands-on individual. I myself have been caught in this trap. Most of these instances occurred when crew members were injured at incidents. However, it is at these instances when it is most important that someone stand back and focus on what needs to be done.

Even at single-unit responses where informal command is used, Command needs to resist the temptation to participate in the incident other than to direct it. At an EMS incident, even if only two members are on the scene, the "informal incident commander" must be the one who focuses on the whole. This allows the second member to take vitals and administer treatment. There are (and will be) times when the person in charge (the informal commander) must participate in treatment to save a life. My advice is that if this becomes necessary, call for additional units to regain total control of the incident. More about applying the incident management system at an EMS incident will be discussed in the last section of this text.

INCIDENT PRIORITIES

As stated in Chapter 1, Command has four priorities at every incident:
- firefighter safety,
- civilian safety,
- stop the problem, and
- conserve property.

With those priorities in mind, Command must make logical assignments based on the following:

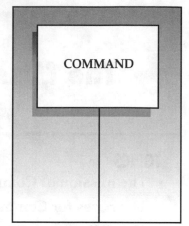

Figure 10-1. Command

- the situation: using a size-up process that involves looking at a picture,
- Command's priorities (as listed above), and
- the availability of personnel and apparatus.

As the situation changes, so does the priority of the assignments. What is essential now might have been third or fourth on the list several minutes ago.

SIZE-UP VS. PREPLANNING

Some basics must be discussed here. Some texts say the size-up should begin at the receipt of the alarm. While I agree in part that you should begin to plan for the incident the moment the alarm is recieved or even before (in the case of form preplans), I believe the first view of the incident is the most important factor in detemining what needs to be done first at an incident. Another way of ending that last sentence, and perhaps more important to the outcome of the incident, would be ... most important in determining what the first on-scene personnel activities will be. After determining what the conditions are (smoke, open flame, for example), the specifics of the structure (occupancy, construction type, and so on) come into play. However, if the situation does not affect or will not be influenced by the specifics of the structure or occupancy type, the specifics are not significant. However, the converse is also true. If the situation affects or will be influenced by the specifics of the structure, then the specifics must be considered. However, the situation is the first step.

SCENARIO 1

A fire call is dispatched to Industrial Heat treating plant in the center of the city. The potential exists for fire, explosion, release of toxic/caustic materials, and so on. If the first unit arrives *and no fire or products of combustion are showing* (the situation), the preplan (the specifics) of the occupancy for the most part is unimportant.

However, if you observe smoke, fire, or occupants (workers) running from the structure on your arrival at the scene, you must indeed consider the specifics, and quickly.

It is not always possible to do what we would like to do because of a lack of personnel. Initial attacks with backup may not be feasible. It may be more prudent to assign a backup crew to hold the stairs in an apartment fire with occupants still fleeing the structure while we are trying to enter.

Figure 10-2. The officer sizing up must look at the picture in front of him to determine the incident's to-do list. What sectors would be on your to-do list at this incident?

Size-up—the officer rapidly evaluates the picture in front of him to determine the incident activities (or to-do list)—will include the following at a fire:
- the extent of fire and smoke on arrival,
- structure type (if a structure is involved),
- occupancy,
- exposure potential, and
- actions of occupants and bystanders.

Extent of Fire and Smoke on Arrival

In my mind, the picture in front of the officer is the most significant factor in determining the incident's needs. In my classes, I tell my students: "Look at the picture. What you see will give you a to-do list." Notice that these tasks are referred to as a to-do list instead of as Command's priorities (discussed earlier in the section and in the second chapter). It is important to keep Command's priorities separate from the to-do list. Command's priorities remain constant. The to-do list can change not only from incident to incident but from minute to minute.

Figure 10-3. A taxpayer with residences with no evidence of a fire. (Photo by author.)

In Figure 10-3, the taxpayer/residential structure shows no evidence of a fire. No smoke or flame is visible. The to-do list at this incident is pretty simple:
- determine why you were dispatched to this incident, and
- handle any emergency that arises.

It would be expected that you will go mobile and go up to the house and ask the occupants if they have a fire. Chances are that if there is a fire, it is small and the needs of the incident will be few. If you find a small fire, then you should handle the situation as procedure and accepted strategic concepts dictate. If there is no sign of a fire and no one is at home, then it may be prudent to check a callback number if available. If no visible signs of fire are present through windows or other openings, then this probably was either a malicious false call or a good-intent call, and you can go back without forcing entry.

Figure 10-4 shows a single-family structure that is well-involved. Command's to-do list at this incident would include the following:
- exposure protection,
- attack in the original structure,
- vent,
- search, and
- backup.

Figure 10-4. A single-family residence that is well-involved. Notice the 1½-inch initial attack line. What size line would you pull at this fire for exposure protection?

All of these assignments would be made to accomplish the "list," bearing in mind the priorities of Command:

- *Firefighter safety:* Commit necessary crews to protect Exposure 2. This can be done relatively safely from the intersection of Sides 1 and 2. Send a crew inside Exposure 2 to ensure that no fire got into the structure. Consider firefighter safety concerning the original fire building. Take lines from the unburned portion of the structure and advance them toward the burned section of the structure—in this case, from the rear of the building, probably stretched along Side 4.
- *Civilian safety:* Once it has been determined that crews can safely enter and operate inside the structure, civilians must be accounted for. Veteran firefighters should be able to recognize when fire conditions will allow occupants to be rescued. With increased response times, larger districts, downsized departments, and quicker periods from incipient fire to flashover, firefighters' ability to rescue and remove trapped or unconscious victims has been reduced. If the severity of the fire and its by-products have apparently "won" the battle at a fire, personnel must not be injured removing *things.* Civilian safety is paramount, but there is a direct correlation between our ability to remove living people from structure fires and the extent of fire and its by-products on arrival.

An aspect of civilian safety that often may be overlooked is the safety of onlookers and displaced occupants. Their safety may be better protected if scene tape is used to establish an exclusion zone that only fire personnel can enter. Liken it to haz-mat zones:

   —hot zone—inside the structure or area of smoke: Members in this zone must have SCBA on and in service.
   —warm zone—inside the exterior exclusion zone (taped-off area); only fire personnel.
   —cold zone—outside the exclusion zone (taped-off area): civilians, media, police, occupants, and so on.

- *Stop the problem.* At this fire, this is best accomplished by the sequence confine, control, extinguish. The first step in confining a structure fire is to look at the picture and ask, Where do I believe we can actually stop the fire? In some instances, it is realistic to attempt to confine the fire in part of the room of origin (as in a rubbish fire), the room of origin (as in a mattress fire or one involving food on the stove that has extended to the cupboards), the floor of origin, or the structure of origin. There have been fires where Command would have felt good if the fire could have been

Figure 10-5. *Sometimes you feel good if you can confine a fire to the block of the fire building.*

stopped three houses down from the original fire building or even at the block of the fire—in essence, exposure protection. Whether there are interior (as in a specific room or floor) or exterior exposures, the fire must be controlled by placing adequate hose streams or using other methods of stopping the production of heat and the advance of the fire and moving it back toward the area of origin—the process of darkening down the fire. In Figure 10-4, it would be realistic for Command to try to keep the fire confined to these two structures With the amount of fire impinging on Exposure 2 and the size of the initial exposure line pulled, I would expect at least minor damage to Exposure 2. However, I would expect that both buildings would be standing when we leave.

Finally, all hot spots should be extinguished through overhaul.

- *Conserve property.* As much property as practical must be conserved in the extinguishment and overhaul processes. Salvage is the process of protecting the occupants' belongings (including as much of the structure as possible). Tarps can be spread, fans used, and so on.

Structure Type

After the extent of fire (or other by-products of combustion) has been evaluated, the structure type (if the incident involved a structure fire) must be considered.

Structure types may be classified several ways. One prominent one is the NFPA rating, Class I (basically fire resistant) to Class V (basically combustible wood frame). This method is most beneficial for writing up inspection and fire reports. For fireground officers, I believe, knowing the materials used in construction more aptly helps in determining the cause-effect relationship of fire to building materials. Types such as wood, ordinary, concrete, steel, and combinations of the above are prevalent across the United States. Specific texts on the relationship of building construction to the fire service are available. Specifics on the above-mentioned construction methods, materials, and their reactions in fire are beyond the scope of this text.

After stating that, remember that you must consider the specific structure type in the size-up process. Fire in a single-family balloon-frame structure will propagate and spread differently than would a fire in a concrete and steel high-rise structure.

Occupancy

Occupancy types range from residential to storage and business occupancies. As it relates to occupancy type, size-up is concerned with the following:
- What is the building being used for—school, restaurant, house, and so on?
- How many people could be in the structure at this particular time?
- Is there any special process that would affect escape and fire spread?

Exposure Potential

Where is the fire going, and what can it affect? There are two types of exposures:
- Interior exposures: the spread of fire within a structure. It can be from part of a room or from floor to floor. Internal exposures are protected best by two actions:
  —cutting off the spread of fire by the proper placement of attack lines and nozzle applications, and
  —checking for extension as quickly and as thoroughly as possible.
- Exterior exposures: the spread of fire from one structure to another structure or other object of value (such as a vehicle or telephone pole). External exposures, discussed at length later in this section, are best protected by
  —covering with a film of water to keep the surface of the exposed material below its ignition temperature,
  —gaining access to the exposed structure (if applicable) to make sure the surface application of water is sufficient, and

—decreasing the number of Btus produced by the fire.

The potential for fire spread to external exposures must be evaluated as soon as possible. Under normal circumstances, if a fire has spread to another structure, first lines probably should be used to protect the exposures. Additional lines will be needed for extinguishment.

The Actions of Occupants and Bystanders

One of the first indications of the severity of the incident (other than visible smoke or fire) will be the actions of the occupants and bystanders. If someone is in the street waving responders down as they approach and others are standing and pointing, then some sort of incident probably is in progress. This may have little bearing on the incident itself or its severity (what is a big fire to a civilian may be a booster fire to a firefighter), but it should serve to get Command's attention and start him looking in the right direction to get his mental processes flowing.

UPDATING AND EVALUATING

After Command makes his initial assignments, based on his to-do list, he waits. At this time, he should look for obvious changes in the fire and its byproducts from the outside. Again, Command must make every effort to remain at the command post. He should be looking for the following:
- *Changes in the color of the smoke:* Light, less forcible smoke most probably means Attack has found the seat of the fire and is hitting it. Smoke that is getting progressively darker may mean that the fire is spreading or reaching flashover range.
- *Changes in the force with which the smoke is leaving the structure.* The rule is, the more pressure, the more fire. If the pressure lessens, chances are that the fire is being knocked down or running out of fuel. If the pressure of the smoke is increasing, it may mean that the fire has entered the structure itself.
- *Changes in the location of the fire.* If the fire bursts out into locations remote from the original fire site, it has had multiple beginnings (areas of ignition), or the fire has entered the structure and is moving undetected in void spaces.

Command should also rely on input from his sector officers. As stated in Chapter 8, one of the three responsibilities of every sector officer is to keep Command informed. In Toledo, it is required that benchmarks be used to do this. If Attack tells Command that the fire has been knocked down on Division 1 and that he is checking for extension, and Command observes heavy fire in the attic, then Command knows that the fire has entered the structure and fire may be elsewhere.

Once initial assignments have been made, Command can and should update his to-do list periodically. What may have been number 1 on Command's to-do list may not even be a concern five minutes later. I like to have at least one staged unit (normally an engine company) for additional assignments until the fire is under control. If it is not used by that time, I normally send it back.[1] As it relates to using and returning dispatched units, it is Command's responsibility to do the following:

- gather the necessary personnel and equipment to handle the situation, and
- get units dispatched to the scene back in service as soon as practical.

DIRECTING SECTOR OFFICERS

Command is responsible for the outcome of the incident. As such, he must take responsibility for the actions of the sector officers working for him at the incident. This leads to the question, Where does Command's ability to direct sector officers at an incident infringe on micromanaging the incident?

The answer is situational. It lies in whether Command has the time to personnally direct sector officers. As you will see as you read the next several chapters, any individual promoted to the rank of company officer (regardless of the title) should possess the knowledge, skills, and abilities to function as the officer of fire units. As such, if Command has the time to give direction to individual sector officers, then I'm okay with that. Direction such as: "You're Topside Ventilation," or "Attack, take a line in through the front door," is acceptable. It gives additional focus to the sector officer. I draw the line at much more direction above this. I don't believe Command needs to list tools or cite specific evolutions or tactics that should be used except in extreme situations. That's why we have procedure and training manuals. (Those extreme situations may include specific tactics at haz-mat incidents or at incidents where the officer is new or a rookie is the acting officer.) Finally, if command knows a sector officer is going to or has begun to do something that will be detrimental to the outcome of the incident or that will cause harm to himself, his crew, or others, then Command must stop those actions.

A LAST WORD ON COMMAND'S MISSION

I'm not sure whether this item belongs in the mission statement, but I know it certainly is a requirement for a successful chief officer or company officer. Most structure fires (and even other emergencies such as car accidents and haz-mat incidents) affect two different and distinct things. At a structure fire, the structure is certainly affected by the fire, but so is the owner/occupant. However, many chief or company officers neglect the owner (or occupant, if not the owner).

Most of us talk to the owner/occupant to get information concerning who owns the structure, phone numbers, what may have caused the fire, and whether smoke detectors operated. But many of us neglect their needs. They may have many questions. We go to hundreds of fires a year. The occupant may have only one fire in a lifetime. We can accomplish much with a few minutes of kind words, comfort, understanding, and effort.

Endnotes

1. In extreme cold or hot weather conditions, I may keep that staged unit to help pick up. Normally, interior crews will have taken a beating at fires in these extremes, and the help in picking up goes a long way toward boosting morale.

REVIEW QUESTIONS

1. What is the mission of Command?

2. On what does Command base logical assignments?

3. Define size-up.

4. What are the components of a size-up at a fire?

5. What are the three steps in extinguishment?

DISCUSSION QUESTIONS

1. Discuss incidents to which you have responded or are aware of where the chief or officer in charge (Command) got caught being a "hands-on guy." Did those actions affect the outcome of the incident?

2. Discuss the statement: "The first view of the incident is the most important factor in determining what needs to be done first at an incident."

3. Discuss the "actions of occupants or bystanders" you have encountered. Were your first observations of these individuals indicative of the incident? Explain.

11

The Mission of Attack

TOPICS
- **Mission of Attack**
- **The responsibilities of the attack officer**
- **Attack strategies**
- **Offensive fire strategy**
- **Defensive strategy**
- **Offensive attack methods**
- **Defensive attack methods**
- **Scenarios in which attack methods would be used**
- **The relationship between the attack officer and Command**
- **The basic attack rules of thumb**

KEY CONCEPTS
- ✓ Defensive strategy
- ✓ Direct attack
- ✓ Fog attack
- ✓ 4-NAP fog nozzle
- ✓ Indirect attack
- ✓ Microns
- ✓ Nozzle discipline
- ✓ Offensive strategy
- ✓ Shutoff
- ✓ Tailboard critiques
- ✓ Tip

*The mission of Attack is to **coordinate** fire suppression efforts by tactically placing and directing attack lines to seek out and extinguish all fire in the area assigned and to provide feedback to Command (and other assigned sectors).*

First and above all, the last person whose hands Command wants on the bail of a nozzle is the attack sector officer. Notice that, as in the mission of Command, the word *coordinate* is stressed. Just as Command's job is to focus on the whole incident, Attack's job is to focus on attacking the fire. I do not believe an officer can effectively coordinate a fire attack while crawling on his hands and knees with a nozzle in his hand. Command expects the attack officer to take his crew inside, locate the seat of the fire, and knock down the fire. While the crew is getting hit in the face with steam and smoke, Command expects the

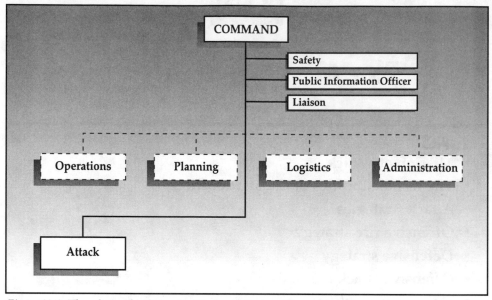

Figure 11-1. The relationship between Attack and Command.

attack officer to be behind his crew (or, on occasion, to move ahead to the next room), looking out for them and thinking of whatever needs to be done next.

I hope you are picking up a pattern here:

- Command focuses on the whole incident.
- Sector officers "tunnel in" (focus) on the task assigned.
- Crews do the work.

Reminder: Officers can't concentrate on an entire fire attack if they are turning over a sofa to hit hot spots or applying an indirect stream to the ceiling, trying to catch the right angle from which to let water rain down on a burning chair.

Some of you may be saying, "We don't have enough people not to let the attack officer be on the nozzle." There are times when personnel may be short. In some departments, that may be all the time. In Toledo, we sometimes use three-person engine companies. There is not a whole lot that a three-person engine company can do. The answer is really very simple: Do only what you can, and then do the best you can. A three-person engine crew cannot take a hydrant *and* pull and stretch an attack line at the same time. The first-in officer, prior to taking command, must decide by looking at the picture what the incident's to-do list is. If it is to quickly take an interior line inside and attempt a quick knockdown, then he cannot afford to leave a firefighter at the hydrant. Most engines carry at least 500 gallons of water in their booster tanks. You can put out a lot of fire with 500 gallons of water if you use a little nozzle discipline:

- Don't throw water at smoke, only at real fire.
- For interior attacks in single-family homes, use an indirect attack

(straight stream off the ceiling) for the best use of water (heat absorption) while providing the least disruption of the thermal balance. (We use this method often.)

- Quickly pull ceilings to check for vertical extension.
- Don't use positive-pressure ventilation until you're sure you have identified the seat of the fire.

Once command has been established, water supplies can be established and backup lines put in place. On the other hand, if Attack knows a lot of water will be needed because the fire is already well advanced, he should leave a firefighter at the hydrant to establish a water supply and assign incoming crews to attack.

RESPONSIBILITIES OF ATTACK

The attack officer has three responsibilities:
- safeguarding his crew,
- directing his crew in attacking and extinguishing the fire in his assigned area, and
- keeping Command informed.

These responsibilities must always be kept in mind. All other actions must be taken after mentally reviewing these three actions.

As indicated in Attack's mission statement, the attack sector officer's responsibilities are to lead crews into the structure (if an interior attack is to be mounted) and to direct extinguishment efforts.

FIRE ATTACK CONCEPTS

Before moving forward, let's discuss a few basic fire attack concepts. Methods of attacking fires have been written about for more than a hundred years. Since entering the fire service in 1975, I have read dozens of texts and articles discussing methods of fire attack. They all have several things in common:

- They differentiate between interior and exterior attacks.
- They differentiate between nozzle and stream types.
- They differentiate between the different types and stages of fire growth.

There are two basic attack strategies and three types of fire attacks (tactics) for each strategy. The strategies have been called by several different names, as have the methods of attack. If the terms I use are not the same as those used in your department, substitute your terms. The name may change, but the basic concept will be the same.

TWO BASIC ATTACK STRATEGIES

The two basic strategies of fire attack are offensive and defensive. Each form of attack is distinct and is dependent on the following:

- the extent of fire on arrival,

- the likelihood of *savable* victims,
- the ability of crews to operate inside safely,
- the stability of the structure, and
- the ratio of risk to the gain.

The objectives of any attack strategy are the following:

- to *confine* the fire (to the smallest area possible by preventing it from spreading),
- to *control* the fire (reduce the production of heat by cooling the area around the fuel to below the fuel's ignition temperature), and
- to *extinguish* the fire (overhaul the fire area to make sure all traces of the fire have been extinguished).

Offensive Strategy

An offensive strategy calls for an aggressive interior attack at the seat of the fire. The key to this strategy is to confine the fire to the smallest possible area. In the offensive strategy, Command believes that an interior attack can be successful, the structure is worth saving, and the risks associated with sending crews inside will not be greater than the materials being saved. I would like to have a dollar for every time I, as a line firefighter or officer, basically risked my life to save the "stuff" we would be shoveling out of windows 15 minutes later. This is not to say that firefighters should not risk their lives to save property. Our job is to save life and property from the effects of fire. However, no firefighter's life or well-being should be risked on property that has already been damaged by fire. There is a fine line here and, hopefully, fire administrators and line officers are learning how to better differentiate between when to and when not to commit to interior firefighting. Property can be replaced.

The Three Tactics of an Offensive Attack

The three tactics for an offensive strategy are the following:

- *Direct attack: Incipient or free-burning-stage fire; a straight-bore or combination nozzle set on "straight stream."*

DISCUSSION: The nozzle is basically set on a straight stream. {As a side note, there seems to be a trend to revert to straight-bore nozzles, which cannot be adjusted between fog and straight stream. They have a shutoff and a straight bore that go from the hose coupling diameter (1½ or 1¾ inches) down to a ½- or ⅝-inch tip.} In my experience, I have found the direct attack to be the most common form of attack. The stream is directed toward the flame (generally at or near the base) and cools the surface of the material burning.

—*Advantages*:
   – It provides greater reach. When the heat produced keeps firefighters at a distance from the fire, then the reach is good.
   – It does not produce a lot of steam. In an enclosed area, that is an important factor for the comfort of firefighters and any unknown victims.

Figure 11-2. Direct attack application at the seat of the fire.

– It probably will not drive out other crews working in the area.
– It gives excellent penetration for deep-seated fires.
– It provides a great deal of knockdown power and quickly darkens down a large quantity of open flame.
– It does not seem to push the fire away from the nozzle as easily as a fog stream.

—*Disadvantages:*

– More water is needed ounce-for-ounce in relation to the amount of fire than with the other forms of offensive attack. The reason for this is that the water droplets should be as small as possible for the maximum absorption of heat. When a straight stream is used in a direct attack, the water droplets are quite large.
– There is more water damage.
– The direct attack disturbs the thermal balance, diminishing visibility. When you have good open-burning fire and then hit the base of the flame with a straight stream (direct attack), "the lights go out."

• *Fog attack: Fire at free-burning stage; a fog or combination nozzle set on at least a 30° fog pattern.*

DISCUSSION: By adjusting the stream on a combination nozzle, you can open or widen the nozzle stream up to a 60°-wide pattern, creating a fog stream. Some nozzles are designed specifically for fog streams. Fog streams certainly have their purpose, but, in my opinion, they should not be used in structural firefighting. I have worked with older officers who would actually become violent if they saw you use a fog stream inside a structure. That always seemed to me to be a bit overdramatic. I would advocate the use of a fog stream inside a structure only in two instances. The first is when a fire

Figure 11-3. Fog attack directed at burning furniture.

must be "held" in a specific area. A fog stream may hold fire to a hallway or other area while crews retreat. This action should be used only in an emergency. The second is for quick ventilation.[1]

— *Advantages:*

–This method provides maximum heat absorption. There is a direct correlation between the amount of heat absorbed and the size of the water droplet. (The optimum size of a water droplet is approximately 350 microns.) Most firefighters do not carry measuring devices, so we will say "the smaller the better."

–A fog stream can hold or even change the direction of a fire.

— *Disadvantages:*

–Maximum absorption yields maximum production of steam. The human body cannot tolerate moist heat nearly as well as dry heat. If you have ever been steamed out of a building by some idiot using a fog line from the outside, you know exactly what I mean and how quickly the atmosphere can change when a fog stream is injected into a heated atmosphere. Even Nomex® hoods afford little protection against the generation of steam. If savable victims are inside a structure, imagine what a fog stream could be pushing onto them.

–Of the three methods of offensive attack, fog streams disrupt the thermal balance the most. If a fog stream is injected into a room where there is free burning and only light smoke with good visibility, the visibility will quickly be diminished. If you must see what you are doing or are looking for (as in a search), you probably do not want to use a fog stream.

–There is little reach and penetration with a fog stream. The majority of the stream evaporates or is pushed away from the seat of the fire by convection currents.

• *Indirect attack: Incipient or free-burning-stage fire; a straight-bore or combination nozzle set on "straight stream."*

Figure 11-4. Indirect attack. Straight stream deflected off the ceiling to break up the stream, allowing it to "rain" down on the fire.

DISCUSSION: With the indirect attack, a straight stream is aimed at the ceiling and then deflected off the ceiling and allowed to rain down on the fire. While approaching the fire area, you aim the stream toward the ceiling at an angle that permits the deflected water to hit the burning surface. After using this method hundreds of times at the training tower in the burn building and at actual fires as a line officer, I believe it to be the best method of interior fire attack for most fires.

—*Advantages:*
  – The straight stream produces little if any steam. The droplets deflected off the ceiling are big enough and, therefore, do not overproduce steam and vaporize (turn to steam) before they reach the seat of the fire.
  – The reach of a straight stream makes it possible for firefighters to stay back from a free-burning fire.
  – Of the three methods, the indirect attack maintains the best visibility and disrupts the thermal balance the least.

—*Disadvantages:*
  – More burning material needs to be overhauled than in a direct attack.
  – You do not get the penetration of a straight stream.

Table 11-1. OFFENSIVE ATTACK METHODS

Attack Type	Advantages	Disadvantages	Comments
Direct Attack	Reach, penetration, low steam production. Won't fan or push fire.	Disrupts thermal balance, more water required (more water damage).	Best for incipient fires and those in flashover.
Fog Attack	Can hold fire and heat to a specific area. Can be used for ventilation. Best heat absorption.	Produces great amounts of steam. Can kill civilians and injure firefighters. Disrupts thermal balance. Little reach.	Use only in an emergency. Will push fire to other areas of the building.
Indirect Attack	Good reach. Little disruption of the thermal balance. Better heat absorption. Less water damage.	Little penetration. Overhaul greater.	Best for free-burning fires, such as furniture that is well involved.

Defensive Strategy

The defensive strategy is an exterior attack on the fire used when Command believes no savable victims are in the building and the amount of fire involvement or questionable stability of the structure makes it imprudent to send firefighters inside. Defensive attacks are used in the following situations:

- The fire has reached flashover and *at least* one floor is totally involved.
- The structure is vacant and well involved in fire, heavy smoke conditions exist, or there have been previous fires in the structure.

Defensive attacks are holding attacks. As mentioned earlier, Command must, in his size-up, ask himself, Where do I want to stop the fire? It may be on a specific floor, in a specific structure, or on the block on which the structure is located.

Tactics of a Defensive Strategy

Whichever of the three tactics is used in a defensive strategy depends on the following factors:

- the amount of fire in the structure,
- the stage of burning, and
- the stability of the structure.

The three tactics are as follows:

- *Direct attack.* The direct attack in a defensive strategy is much like the direct attack in an offensive strategy. The nozzle is set on a straight stream and directed onto the burning material. This method is used when the fire is in flashover or the free-burning stage (visible fire) and firefighters can't get close to the structure because the structure's stability is questionable or a great degree of radiant heat is being generated. This is the type of attack most people think of when they think of large fires; firefighters stand outside and dump water into the building with aerials, platforms, monitors, and 2½-inch handlines. After the fire has been darkened down, Command (and Operations) determines whether an interior overhaul will be made or the "dozer" will knock down the structure.

Figure 11-5. A direct defensive attack on a structure that has no survivable victims and questionable stability.

If Command calls for a defensive stategy at a fire, *no crews should enter the fire structure for any reason without (1) notifying Command (or Operations) and (2) having Command (or Operations) approve and confirm their actions.* Command (or Operations) has the idea in his head that all savable people, including firefighters, are out of the structure. He will then make certain strategic moves based on that fact. At that point, he will not get overly concerned if a portion of the structure collapses or breaks out in fire because *no one should be in there.* But, say you are in Sector 3 and and you let your crew convince you (or you may even convince yourself) that "you can get a great shot at the fire from just a few feet inside the door." You are putting yourself at great risk. Command and Operations may never even blink an eye at a collapse or a new area of involvement if they believe no one is near that area.

In the direct attack, straight streams are directed toward the fire through open windows, ventilation holes, collapsed areas, and so on. It is important that firefighters maintain their distance and that the safety officer mark off collapse zones (using physical or imaginary boundaries).

• *Indirect attack.* The indirect attack has only one application. It is used

Figure 11-6. A firefighter using a defensive direct attack from a porch.

when Command believes there is a potential for backdraft (in the high-temperature, smoldering stage). Conditions indicative of a backdraft are the pushing and pulling of smoke from inside a structure, the absence of visible flame, the presence of an oily residue (condensation) built up on windows, and doors that are hot to the touch. These conditions occur when a fire has been burning for some time and has consumed the oxygen in the area (or structure) to a level below 16 percent while retaining the heat and the fuel. If air (oxygen) is introduced into the area, the fire will now have the third item necessary for combustion. With the high heat and vaporized fuel, the improper injection of oxygen into the atmosphere could lead to a violent explosion (backdraft).

Although I have been to many fires in my career, I was involved in only one backdraft fire and heard of a few others in Toledo. The fire in which I was directly involved is as vivid to me today as if it happened yesterday.

SCENARIO

It was a cold night. I was the officer of #7 Ladder. When we arrived at the scene, it seemed as though everyone was walking around in slow motion (this was before the incident management system). In my mind, I knew that the chief felt the building had the potential for a backdraft. Even though no orders were given, I had my crew ladder the roof of this two-story storefront to open it up. At this time, no fire was showing; just slight smoke from the eaves. No one was kicking in doors or breaking windows, as we usually did.

We put a 35-footer up to Side 4 and were going up to punch a hole in the roof. I was the first man up. Tim Boaden was about eight feet up the ladder, and Mike Palechety was butting the ladder. Just as I stepped off the ladder onto the roof, someone (I never found out who) opened up a big picture window. The explosion threw me up about a foot in the air. It blew Tim off the ladder and an engine driver, Don Dunbar, under his engine. He landed up against the curb. He never returned to work after that day. Tim and I were not hurt.

To decrease the potential for a backdraft, two things must be done: First, a vent hole must be made near the top of the structure (hot gases rise) to safely carry out the products of combustion. Second, the inside atmosphere must be cooled by injecting steam (or fog) into the lower area of the structure. The heat and draft from the ventilation hole will force the hot gases and steam out of the top of the structure. Additionally, the water injected into the super-hot atmosphere will turn to steam, which will expand approximately 1,600 times. This influx of steam will also force steam into hidden areas of the structure.

To be effective, an indirect fire attack must follow this sequence:
- Identify the conditions that indicate a potential backdraft.
- Stop all efforts to enter the structure.
- Create a ventilation hole in the roof, at the ridgeboard or highest portion of a flat roof. When the hole is opened, smoke and hot gases should be coming from the vent hole. A torch effect of heated gases igniting as they reach the oxygen-sufficient air outside this vent hole may also be observed. This is normal, and no effort should be made to extinguish this flame. It shouldn't be hurting anyone or anything.

After the vent hole has been opened and the ventilation crew has left the roof, a small hole can be opened in the lowest portion of a door panel, and a 60° fog stream can be injected through the hole.

Then, wait.

After about five to 10 minutes, the ventilation team should return to the roof and check the heat coming out of the ventilation hole. If little detectable heat is coming from the hole, it is safe to enter the structure. The procedure for safely entering the structure is as follows:

Open the front door from the side, and wait for a few minutes before entering. If nothing happens, enter with care. Send in only a small attack company to check for hot spots and to make sure there are no other potential hot spots. Carefully open every door you encounter *after feeling each one for heat*. Do not open any door that is hot to the touch. Instead, punch a small hole in the lower portion of a panel and inject a 60° fog stream through the hole (as described above). Usually only one vent hole is required. Once you are inside and all areas have been opened, extensive overhaul will probably be needed.
- *The combination attack.* This approach is used when an area of the structure has reached the flashover stage and the structure is *approachable*. This is the fire at which you pull up to a vacant house on fire that has an area accessible and in flashover. You decide that it would be best to knock

down the fire from the outside and then let it sit for a few minutes until you can safely enter for overhaul. I have used this method several times as a line officer, and it works. This method can also be used on an occupied structure that is in the flashover stage on the first floor. It may provide a quick exterior (defensive) knockdown, prior to beginning an offensive interior attack. Remember, however, that using this method in an occupied house will push heat, smoke, and fire to uninvolved portions of the structure.

In the combination attack, you approach an open window or door with the nozzle set at between 30° and 60° fog. Move your hands back from the nozzle and begin to rotate the nozzle in a clockwise[2] motion. This will create a turbulence and have a dramatic cooling effect on the fire. If done correctly, this method will provide a quick knockdown with suitable protection for the nozzleman.

Note: A hose stream will put out all the fire it can in about 30 seconds. With any method, if the fire is not darkened down in about 30 seconds, you don't have enough water. The fire is putting out more Btus than your hose stream can absorb. Get another or a larger line.

One last word of caution: Never attempt a defensive attack if

- savable victims may be inside the structure, or
- crews are inside the structure.

Command must know the locations of on-scene crews. If Command decides that a defensive attack is warranted, he must ensure that all companies are outside prior to the opening of any exterior nozzles. A head count and check-in system must be part of every department's SOPs. A newer emergency accounting of on-scene crews is called the "personnel accountability report" or PAR. Many departments now use this system to get a quick accounting of crew members. Command or Operations calls for "emergency radio traffic" and then says, "Command to all units: Give me a PAR." When an officer physically accounts for all of his members, he reports this to Command or Operations by saying, "Command (or Operations): Attack has PAR." Command or Operations should know the identity of the attack and other sector crews.

THE RELATIONSHIP BETWEEN ATTACK AND COMMAND

As stated earlier (again, I hope it's becoming a familiar theme), Command's responsibility is to focus on the whole. He should not be micromanaging the attack. If Command finds himself focused on the attack, then other aspects of the incident may be passing him by. The attack officer should be focusing on the attack. His task is to seek out and extinguish fire. That's it!

One question asked of me is, If Command has a specific method of attack or an area from which he believes the attack should be mounted, should he tell the attack officer? Absolutely!

When I begin to make the assignments on my to-do list, my first consider-

Table 11-2. DEFENSIVE ATTACK METHODS

Attack Type	Advantages	Disadvantages	Comments
Direct Attack	Reach and penetration.	Not great heat absorption.	Used for reach at buildings that are in danger of collapsing.
Indirect Attack	Used for high-temperature, smoldering fires.	Not too many other practical applications.	Could be used at shipboard fires if conditions are right.
Combination	Allows for knock-down of fire in flashover. Great heat absorption.	No penetration. Extensive overhaul needed.	You must get in after knockdown to overhaul, or rekindle will occur. Can push fire to other areas.

ation is, Can we mount an offensive attack, or will we have to go to a defensive operation first? Most of our fires are offensive fires. A defensive attack is rarely mounted initially. Most times, companies are picturing an offensive attack while responding. I don't usually announce that we are conducting an offensive attack. However, I always make sure that all responding units understand when I want an initial defensive attack. I usually announce and reannounce that fact throughout the first several minutes of the incident. *If I announce that we are going to mount a defensive attack, I never want to find any personnel inside that building without my approval.*

In the offensive mode, I normally have a picture of the location from which I want the lines to be taken in. If I am on the scene prior to the attack company and have the time to do so, I will tell the attack officer my preferred location for taking in the lines and where I think the fire is.

Command to Engine 5: You'll be Attack. It looks like the fire's in the front room on Division 1. Take a 1¾-inch line in the back door, and push it out the front. The window is vented.

This leaves little room for the officer on Engine 5 to mess up. It's hard to criticize someone for not doing whatever we were *thinking* they should be doing. If you have the time, tell them. Once they're inside, they're basically on their own. Some rules of thumb that apply to attack are mentioned at the end of this chapter.

If we have taught our officers and have provided feedback after previous fires (as in the form of "tailboard" critiques), then they have an idea of what it is we want them to do. Communication is the major key here. If they have a question while inside, they should ask. If conditions inside change, they should tell Command. Remember also that if Command designates an officer to be Attack, that officer is responsible for the fire attack in the entire structure.

Unless Command indicates differently, it is up to Attack to determine (1) how and where to get into the building, (2) where the attack line will be taken, and (3) the method of attack to be used.

If you are assigned to be Attack, approach the structure with your crew; look at the building for visible fire; and observe the volume, force, and color of the smoke. Choose an entrance you believe will give you the best chance of cutting off the spread of the fire. Listen to the radio and the assignments Command makes. The orders he gives provide information about how he perceives conditions, such as in the following examples:

- If Command orders exposures to be protected, it indicates the direction in which the fire is heading and venting itself.
- If Command orders ventilation of the roof, he believes the fire has gotten a substantial hold of the structure itself. It is already or has started to head toward the attic or cockloft.
- If Command orders positive-pressure ventilation, he probably believes the fire is small and has not gotten into hidden spaces yet.[3]
- If Command orders additional attack lines, there's probably more fire than was originally thought, and he *should* confirm Attack's location and then inform Attack that he has assigned another Attack to operate on another division (a floor other than the one the originally assigned Attack is working on).
- If Command orders Attack out, he knows something Attack doesn't. Attack should leave the building, take a head count, and report to Command when outside.

SCENARIO

Let's say that Command assigns you (Engine 5) to be Attack and that you and your crew start inside. You kick in a door, advance a few feet, and determine that you can't handle all the fire in the building effectively. Suggest to Command that he may want to get another attack company on Division 2.

Attack to Command: We've got a lot of fire on Division 1. You might want to get another company started on Division 2.

Command: Okay! Engine 5, you'll now be Division 1 Attack. Engine 6, you're Division 2 Attack!

Engine 5: Okay on Division 1.

Engine 6: Okay on Division 2 Attack.

Now you know your boundaries and responsibilities. Command knows that he has two attack companies. These two companies can talk directly to each other to fight the fire effectively.

Division 1 Attack to Division 2 Attack: The fire seems to be in the walls on Side 3.

Division 2: Okay.

One final note in the relationship between Command and Attack: For lack of a better term, I like to think of Attack as the person in charge inside. He, above all other officers (except maybe the backup officer), should have the best idea of what is going on inside and what else may be needed. If too many members are on a stairway or standing around, someone needs the authority to get them out of the way—or even out of the building. If a line is pulled and does not need to be charged or if the fire has been knocked down prior to the assignment or placement of a backup sector, then Attack should make that call and inform Command (who will have the last word). Basically, if things aren't going well inside and a tactical decision needs to be made, Command wants someone to take control to (1) make the necessary adjustments and then (2) inform him. Because of his position in the majority of fires, that someone should probably be Attack.

BASIC ATTACK RULES OF THUMB

There is more than one right way to fight a fire. One of the ways I assess how an incident has been handled is by asking, Could I defend that action in court? There always seem to be several different (and basically correct) ways to handle an incident. There are also as many, if not more, wrong ways to handle an incident. The final section of this chapter deals with some specific "rules" that have been passed down to me by a variety of sources—from my father to past and present supervisors to texts and articles I have read. I shall paraphrase these concepts. If the individuals who have shared their original concepts with me are reading this, I want to say, "Thank you for giving me these specific rules." I have applied them inside a structure while crawling on my hands and knees with fire over my head and outside while awaiting feedback from interior crews.

- **If fire wants out, let it out.**[4] Avoid the temptation to push fire back inside a structure. Fire will be drawn toward openings. If an overly aggressive nozzleman hits a self-venting fire, he might push fire, heat, and smoke back toward its point of origin. This problem can be caused in several ways:
  —*pushing fire back down ventilation holes.* A good rule to follow is, never let truckies have a hoseline. Time and time again (on television and in videos), I have seen ventilation companies doing this.

—*hitting fire venting out from under eaves and the exterior soffit.* You'll be knocking down exterior fire, but you'll also be pushing fire (by fanning the flames with the turbulence of the hose stream) back inside the structure, causing it to spread vertically and laterally from the original point of venting. If you observe fire venting in this fashion, send an attack or overhaul company to that area and have the crew punch it out from the inside. That way, you will have a much better way of controlling spread and ensuring that you get all the fire.

—*fire venting from a window or door—"moth-to-flame syndrome."* Where are the heat and flame going when they vent out a door or window? Outside where they (1) dissipate, (2) scorch siding, or (3) impinge on or radiate heat on an exposure?

Figure 11-7. If a fire wants out, let it out! Truck crews should resist the temptation to hit fire from above.

Number 1 is a no-brainer. Let it go. It isn't hurting anyone. For number 2, cool the siding with a stream by hitting above the window or door and letting the water flow down the side of the siding until interior crews push the rest of the fire out. (Never direct a stream inside a window or door when interior crews are advancing on the fire.) The siding then can be overhauled if necessary. In the case of number 3, again, cool the exposed surface with a constant stream of water until the production of heat is diminished and the amount of radiant heat put out is not significant. But again, never direct a stream inside a window or door when interior crews are advancing on the line.

• **Attack from the uninvolved area toward the area of involvement.** Another way to say this is, if the fire is in the front, attack from the rear;

and if the fire is in the rear, then attack from the front. This helps in two ways. First, it diminishes the spread of fire throughout the building (this is the "confine" in confine, control, and extinguish). A hose stream (even on straight stream) can push a lot of fire around. It pushes heat and smoke away from the advancing nozzle. Second, if fire is between you and savable victims, you will be pushing heat, flame, and smoke on them. This will certainly diminish their survivability. For example, say the fire is in the living room. The occupant started to exit through the kitchen and passed out from the smoke or lack of oxygen. A line is taken through the front door. The possibility exists that you will be pushing heat, flame, and smoke onto the victim.

If, however, the line had been taken to the rear of the house and through the kitchen, you probably would have stumbled on the victim and would have avoided the possibility of steaming the victim to death.

If fire is on the second floor (or higher) of a two-story house (or higher), I have found that the quickest way into the structure works best—usually that would be the front door. In most homes, stairs to the second floor are off of or connected to the living room.

- **Generally, a hose stream will put out all of the fire it can in about 30 seconds**. If you are at the top of a stairway and have a line directing water at a well-involved second floor or attic and you don't darken down the fire within 30 seconds or so, get more water. I have observed companies (my own and those on television programs) work a hoseline for many minutes at a time and not be able to advance or darken down the fire. In the case of my fires, if the house is vacant, I may use it as a learning experience. After a few minutes, I'll get another line (or a larger one such as a 2½-inch) up with the attack company.

  They're amazed at what a little more water will do. Additionally, I believe that some fires aren't meant to be put out. Plain and simple, some fires are beyond our capability. At large warehouses or in vacant apartment buildings in which the fire is well advanced on our arrival, we may not be able to make even a dent in the progression of heat. I have been at fires where our streams vaporized 20 to 30 feet before reaching their target. Just remember, if a stream doesn't darken down its objective in about 30 seconds, the fire is producing more Btus than the water can absorb. You need more water.

- **It's not always good to put out the fire right away.** When I was a line officer, I used to get so mad at overanxious nozzlemen who would knock down fire as soon as they saw it. Many times, the free burning was not going anywhere and the light being provided by the flame provided visibility for search or surveying the surroundings to determine what should be done next. If the fire is not extinguished correctly, the thermal balance is disrupted and visibility is destroyed.

- **Notify Command prior to changing your floor of operation.** One of the cardinal rules of firefighting is, Keep Command informed. Command expects to see certain things after he makes his assignments. (Additionally, there are certain things he does not expect to see after he makes assignments.) If the fire is on the first and second floors and you knock down all the visible fire on the first floor and then decide to take the line up to the second floor, tell Command.

Attack to Command: We've got the fire knocked down on Division 1 and are going up to Division 2 to check for extension.

Figure 11-8. Attack a fire from the uninvolved area toward the area of involvement. When fire is in the rear, the initial line should go in through the front of the structure.

Command now knows several things and expects to observe others. First, hopefully, he will no longer be seeing any fire on Division 1. (I have been at several fires where Attack thought he had the fire knocked down and did not know that a lot of fire was showing in another area of the first floor.) Second, Command will expect to see smoke conditions change on Division 1. He'll expect to see lighter (whiter) smoke with less force behind it. Third, he should begin to see activity on Division 2. (He now knows that he has crews on Division 2.) Prior to this, had the second floor started to flash over, Command should not have panicked. No one would have been on Division 2.

- **For multiple fires on the same floor, bigger is first.** When you encounter multiple fires on the same floor, you must first decide the location from which the line will be taken in. Work the biggest area of involvement first. Command should have assigned a backup[5] company to "watch your back" until all the fires have been put out. While the crew is hitting the first (and largest) fire, the attack officer should be back from the nozzle, determining where the nozzle should go next.
- **When you encounter fires on multiple floors, start at the lowest level possible without being cut off from your means of egress.** The one area of real concern is a fire where the basement and upper floors are

Figure 11-9. A hose stream will put out all the fire it can in about 30 seconds.

involved. If access to the basement is from a side door, as in many older residences in Midwestern cities, it should be possible to take a line downstairs and have backup right at the stairs holding the exit for you. If the fire is on grade level and upper floors, attack on the lower floor and work your way up to the attic. If backup is doing its job, you should have no problem with stray, unseen fires or flare-ups.

- **Opposing lines—bigger is best.** The problem occurs when one crew takes a line in the front door, another crew freelances and takes a line in the back door, and the fire is in the middle. The rule of thumb here is, Bigger is best! Someone here will be a loser. I've seen it several times prior to the incident command system. Believe me, someone will come out of that building madder than heck! There is no excuse for this. In a single-family structure fire, unless a company freelances, Command will be held responsible for hoselines improperly placed from the exterior. If Command sees a line going where it will end up in direct opposition to another existing line, he is responsible if he does not correct the problem.

- **Under normal circumstances, lines sent in to assist other companies should take the same way into the structure.** This rule should stop any problems involving opposing lines on the same floor. Lines going to different floors should not be taken up alternate stairs if possible. Having said this, try to avoid stretching more than two lines in the same stairway. It makes ascending and descending much tougher.

- **Big fire—big water.** This rule applies to the line size that attack companies usually use at a single-family residence fire. Generally, a 1½-inch line is sufficient for a one-room fire in a single-family residential structure fire. You can get at least 100 gpm from these streams, which should be more than enough to extinguish a fire in a room 20 feet × 15 feet × 10 feet (which is a **large** room). At this writing, the trend is toward 1¾-inch lines. With the proper nozzle (1¾-inch) and the correct nozzle pressure (approximately 200 psi on a 150-foot section), you can get almost 200 gpm out of these nozzles. The problem with the 1¾-inch line is the nozzle reaction exerted. To get the rated (desired) flow, two to three firefighters must hold the line. For large flows, I like the old 2½-inch line. You get plenty of water; and once in place, one firefighter can handle the line if necessary.

- **Switch plans, if necessary.** If you've tried several tactics and conditions don't seem to be

Figure 11-10. Under normal circumstances, lines sent in to assist other companies should take the same way into the structure that those companies took.

Figure 11-11. Big fire, big water!

improving, try something else. Go to plan A (whatever that may be) for a few minutes. If conditions still don't improve, go to plan B.

Endnotes

1. Fog streams can be used as a quick ventilation tool, as discussed in Chapter 14.

2. Tests indicate that a clockwise motion is more effective than a counterclockwise motion. The reason for this is not quite clear.

3. PPV and its application are discussed in Chapter 14.

4. Review the combination method of defensive strategy. This may be the only exception to this rule.

5. The mission of Backup is covered in Chapter 13. For now, just know that a backup line should be positioned to protect interior companies.

REVIEW QUESTIONS

1. What is the mission of Attack?

2. List the three responsibilities of Attack.

3. What are the two basic attack strategies?

4. On what is each form of attack dependent?

5. Define "offensive strategy."

6. What are the objectives of a fire attack?

7. What are the three tactics for an offensive strategy?

8. Define "defensive strategy."

9. When are defensive attacks used?

10. Defensive attacks are _____.

11. The type of tactic used in the defensive strategy is dependent on what factors?

12. What three tactics can be used in a defensive strategy?

13. Never attempt a defensive attack if _____.

14. If fire wants out, _____.

15. Attack from the uninvolved area _____.

16. Generally, a hose stream will put out all of the fire it can _____.

17. Notify Command prior to _____.

18. For multiple fires on the same floor, _____.

19. When fires are on multiple floors, _____.

20. Under normal circumstances, lines sent in to assist other companies should be taken _____.

DISCUSSION QUESTIONS

1. Discuss past fires at which a defensive operation should have been mounted from the beginning of the incident. What were the outcomes of those incidents?

2. Discuss the last time your department conducted a drill on a high-temperature, smoldering fire scenario. Do you believe there is potential for one of these types of fires in your community? Explain your answer.

3. Do you agree or disagree with the author's fourth rule of thumb: "It's not always good to put out the fire right away"? Cite examples to support your answer.

12

The Mission of Search

TOPICS

- **Mission of Search**
- **Responsibilities of the search officer**
- **Search strategies**
- **Primary search**
- **Secondary search**
- **Relationship between Command and the search officer**
- **Factors that influence where to search**
- **Facts relating to whether the search sector should have a hoseline**
- **When to remove a victim from a structure and when to leave him there**
- **The oriented method of search and rescue**

*T*he mission of Search is to **coordinate** the prima-ry search efforts in the area assigned. If no specific area is given, then Search is responsible for search-ing the entire building. The word *coordinate* is emphasized for a specific reason. (If you have been reading and paying attention up to this point, you already know it.) The search offi-cer is the last person Command wants to have crawling on his hands and knees, sweeping under beds and in closets.

KEY CONCEPTS

✓ Fire investigation unit
✓ Left-/right-handed search
✓ Oriented man
✓ Oriented method of search and rescue
✓ Safe haven
✓ "Your wall"

RESPONSIBILITIES OF THE SEARCH OFFICER

The search officer has three responsibilities:
- keeping his crew safe,
- coordinating a primary search of the assigned area, and
- keeping Command informed.

The officer assigned to search cannot meet all three responsibilities if he is on his hands and knees, sweeping across beds. If he is "tunneled in" on the

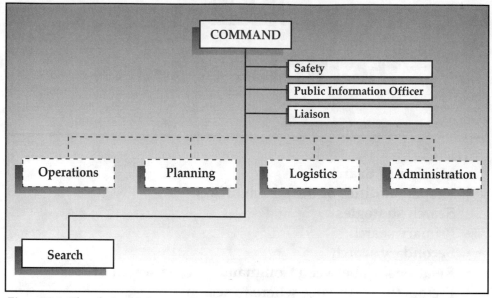

Figure 12-1. The relationship between Search and Command.

search of a specific room, then it's hard to do anything other than search that room. The search officer, or "Search," should be in the area being searched (let's say an upstairs hallway), making sure that all the rooms are being searched (and maybe re-searched) while ensuring that conditions in the area are tolerable and safe for his crew. Additionally, he may need to communicate with other sector officers or Command concerning ventilation and extension of fire. That's hard to do if you are involved in the search.

Personnel issues surface here again. Not all departments run with four-person units (although some run with even more than that). How many fire-fighters does it take to conduct a thorough search in our "bread-and-butter" fires (single-family residential structures)? That depends on the following:

- the levels of training of the searchers,
- the structure type and number of rooms being searched,
- the amount of fire in the structure at the time of the search, and
- the length of time the fire had been burning before the search company arrived.

The sector officer assigned to be Search may have to assist in the actual search efforts under some circumstances, such as that one fire in a thousand at which someone outside is telling us exactly where to find a victim who is still in the structure and there is no time to muster an adequate search crew. This is the fire where you have one chance to get in, get the victim, and get out. With the oriented method of search, introduced later in this chapter, a two-person crew (including the sector officer) can conduct an adequate search in the same time it would take two or three searchers using standard search techniques.

Figure 12-2. A fire crew entering a building to conduct a search.

SEARCH STRATEGIES

Command can use two basic search strategies when making assignments. Both depend on one consideration: Are savable victims in the structure? If the answer is yes, a primary search strategy is needed. If the answer is no, a secondary search will suffice.

The Primary Search

A primary search, usually conducted initially at a structure fire, is a rapid, systematic search of the structure to ensure that all savable victims have been removed from the structure or, in the case of a high-rise or multifamily structure, taken to a safe haven within the structure. It should be conducted whenever there's a chance that savable victims may be in the structure. Depending on the amount of fire involvement on arrival, the occupancy use of the structure (residential, commercial, or vacant, for example), and the information gathered from occupants or neighbors, I put search somewhere on my to-do list and assign a search officer when search reaches the top of the list. (If the condition of the structure at the time of arrival indicates there is no chance that savable victims are in the structure, I assign a secondary search when conditions allow.) If the fire is in a vacant structure that is totally involved on the first floor, assigning search would be relatively low on my to-do list. If, however, the fire is in an occupied bungalow, only light smoke is showing from the rear of the structure, and neighbors tell me they think someone is inside, search probably would be at the top of the list.

Primary searches are rapid searches. A time factor is involved. The human body can be without oxygen (less than 16 percent oxygen content) for only two to six minutes. Depending on the amount of smoke the fire is producing and the response time (including the time it took to notify the fire department), there may not be much time to pull viable victims from the structure. If Command believes that savable victims are inside, he should have the areas in which they most likely would be located searched and even re-searched. It may be advisable to switch searchers and areas after an initial search has been conducted.

I taught search techniques to more than 100 fire recruits in the Toledo Fire Department within a three-year span. During that time, I observed all 100 recruits search in simulated (blacked-out face piece) and actual fire (our burn building) conditions and found that different searchers are excellent at different things. Some are good at sweeping corners and in closets, whereas others are better at hitting the center of the room and under beds. Some always missed the top of the bed. My point is that if you are conducting a primary search and find no victims, send a different searcher into a previously searched room—in essence, switch searchers and rooms. One time in a thousand, that new searcher will stumble on a missed victim.

The Secondary Search

A secondary search is a slow, methodical search to locate deceased victims. After the fire has been declared under control and visibility has been improved by proper ventilation and illumination, a secondary search may be indicated. Crews are sent into the structure to make a systematic search to locate unaccounted-for victims—certainly one of the grimmest tasks on the fireground. The officer assigned as Search in a secondary search must coordinate the operation and keep reminding the searchers of their mission. In a secondary search, beds are turned over or upended, furniture is moved from walls, and closets are thoroughly checked. In essence, no stone is left unturned.

THE RELATIONSHIP BETWEEN SEARCH AND COMMAND

Command's responsibility is to ensure that all things that must be accomplished at an incident are being done.

Command's Assumptions

Once he assigns search, Command must make certain assumptions, such as the following:

- The search officer is actually coordinating a search of the structure or assigned area.
- Search has a specific plan in mind concerning where to search first and last.
- Search will keep the safety of his crew paramount in his mind. (Keep in mind that it is not likely that savable victims will be found in areas of the structure where fire conditions are such that crews cannot operate within those areas.)
- When search changes floors, he will notify Command that the search is being moved to a different division.
- If Search believes that savable victims are in the structure and that he cannot cover the area(s) in time, he will ask Command for more help.

Search's Assumptions

Search also has to make some assumptions:

• If Command receives information pertaining to the search while the search is being conducted, Command will relay that information to Search.

• Backup will be ensuring the safety of the search crew.[1]

As in the case of Attack, if Command designates a company as Search, that company is responsible for searching the entire structure. This is a functional sector or "group" (as in search group), and groups transcend levels and areas of floors. Additionally, if Command does not specify the area in which to commence the search, Search should choose the starting location. Hopefully we have trained our officers to think in terms that will relate to successful searches.

WHERE TO SEARCH

The ability to choose the best location to start a search is acquired through training, experience, and by following case studies.[2] Several indicators can be used to help determine the best areas in which to begin the search. Some of those indicators include the following:

The Time of Day

Nighttime fires will send search crews toward sleeping areas. In two-story homes, that's normally the second floor. With daytime fires, crews will head for living and work areas such as the living or family room, the laundry room, or the kitchen. There may be exceptions, such as the occupant whose work schedule does not coincide with normal shifts.

Figure 12-3. In a daytime fire, search crews may head toward the living areas first.

Figure 12-4. At a fire in the early morning, search crews may head toward the sleeping area first. Note that front porch roofs are excellent for gaining quick access to second-floor sleeping areas.

The Fire Intensity and Products of Combustion on Arrival

The time it takes a residential fire to reach flashover has been estimated to be from between less than four to more than 10 minutes, depending on many variables.[3] One of these variables is "better living through chemistry." More high-Btu-yield materials are found in homes these days. Plastic television cabinets and synthetic fibers covering furniture and floors have heightened the amount of heat produced in residential fires. The more fire and smoke present on arrival, the less likely it is that savable victims will be found. However, if you arrive at a well-involved fire and feel that a savable victim may still be in the structure, conduct a search using the same procedure you would use for a fire attack: Move from the unburned area toward the burned area.

The Location of the Fire at the Time of Arrival

If the fire occurs at night and the location of the fire on arrival is in the sleeping area, the sleeping area and areas immediately adjacent to it must be searched first. If the fire is on the first floor, crews must attempt to make the second floor if at all possible, preferably by interior stairs. There are two reasons for this. First, it is easier to position backup lines on or near stairways to protect search crews if search and backup crews use interior stairs. Second, most victims are found in the location that they were in prior to the fire, such as in bed or in a chair, or on their way to an exit. A human tendency is to flee from the structure the same way you usually enter it. Victims attempting to flee from the second floor usually are found in the upstairs hall, on the stairs or landing, or near bedroom windows. They were trying to get out and just couldn't make it. Victims attempting to flee on the first floor will be found in hallways, near doors, or at windows.

Information from occupants/neighbors.

Occupants may give information that will help you to determine the last known location of a missing victim. Neighbors, for example, can reveal patterns that may pinpoint a potential location—such as noting that an occupant stays up late at night watching television in the kitchen or works the third shift and sleeps in the second-floor bedroom until about 3 p.m. All this information should be noted and relayed to Search. However, recognize that many "goofs" are out there, and it may not always be prudent to take what people tell you as gospel.

Figure 12-5. Could savable people be in this structure?

Figure 12-6. The location of the fire gives an indication of where savable victims will be and, hence, where to search first. Where would savable victims most likely be found at this fire?

Figure 12-7. Information gained from neighbors can help the search effort. Command should pass this information on to Search.

TO HAVE A LINE OR NOT TO HAVE A LINE

Another question frequently asked is, Should the search crew have a hoseline? My answer is, No! That's what backup is for! Let's look at this issue from a personnel point of view. Let's say a four-person crew assigned to search takes a hoseline in with them. How many firefighters does it take to handle a charged (or even uncharged) 1½-inch line? Two, unless you're not going to advance it past the front door. By taking a line, you effectively have cut your search crew (and its capability) in half. Additionally, if a man is stretching and dragging a hoseline, on what is he focused, the line or the search? We know he is focused on the line. Even if he focuses on the line only part of the time, is he giving that victim his best shot at survival?

Let's look at it from the perspective of the fire's intensity. If the intensity is such that the searchers feel comfortable only if they have a line with them, is there any chance that savable victims are in the building? If you don't feel comfortable with 35 pounds of "stuff" on your back to protect you, think about how the victim feels (or felt) without it.

The great thing about the incident management system is that not everyone staged at the incident has a task to accomplish. Backup should be one of the most frequently assigned sectors at a fire. If Backup is doing his job, the search crew has no need for its own line. Everyone has a task, and everyone should focus on his task. That's what makes this system work.

WHAT TO DO IF YOU FIND SOMEONE

Every time I did a search as a line firefighter or officer, in the back of my mind I was hoping for two things:

- that I would be the one to find the victim, and
- that someone else would find the victim.

What a time of uncertainty! As you crawl along, everything you feel seems like a part of a person. Every time you bump into a potential "body part," your adrenalin pumps. Every time you realize it's only a doll or a pillow, you sink back down. Up and down! Up and down!

When I was teaching recruits in 1984, one asked, "What should I do if I find someone?" Everyone in the group laughed except me and the recruit asking the question. After I said, "Why, that's an excellent question," I gave the rest of the group an extra 15 push-ups to do during physical training.

That was an excellent question! I have seen coroners and homicide detectives literally shake because they were so mad when fire crews removed obviously dead victims from the area where the body had been found. Until a definite cause has been established, a fire scene has to be treated as if it were a crime scene, especially if there are fatalities. The fire investigation unit has a difficult time replicating the scene as it was prior to ignition if the victim is removed. This is one of the hardest calls to make at a fire. My rule of thumb is, If there's any doubt, take the victim out.

I have seen "and felt" obviously dead fire victims. Several definitive clues will indicate that a victim has become a fire fatality: (1) rigor mortis sets in a lot quicker due to the heat involved (you pull on an arm to begin to drag the victim out, and the arm does not straighten); (2) the skin color changes to a pale yellow; (3) there is obvious body charring; (4) clothing is burned off; and (5) there are no vital signs. Obviously dead victims should be left where they are found.

I will order that a dead victim be removed if I believe the structure will collapse onto the body prior to removal. You should meet with the authorities who investigate your fires to write specific guidelines concerning protecting the body prior to its removal. Stress to these individuals, if they are not firefighters, that we may still have fire suppression tasks in the area in which the victim may be found; they may have concerns about covering the victim. They may want them covered or not covered, removed or not removed. They have a job to do, as do you.

I do not allow my firefighters to remove victims from a structure after the investigation is over. If the structure has been weakened by fire, I ask for volunteers to put the body in a body bag and move it to safety. But I do not let my members carry the body bag out of the building, past television cameras, onlookers, and family members. First, I do not believe it's our job. The coroner's office or detective normally do that task. Second, I do not believe it sends the proper message. We should not be seen bringing out dead victims. I know as well as anyone that we can't save them all. (I personally had nine fatal fires in 1991. I believe we had a total of 14 that year.) The fire companies at the scene will go to bed that night with enough pictures in their heads. The looks on firefighters' and civilians' faces as the victims are carried out is one of those pictures they need not have, and one that I can keep from them.

THE ORIENTED METHOD OF SEARCH

Many different tactics are used for primary searches. In 1984, I was introduced to the *oriented method of search and rescue* through an article given to me while I was preparing lesson plans for a recruit class.[4] I taught the oriented method to all 63 recruits in that class. The deputy chief of training observed me as I was teaching this method one day and was so impressed by it that he asked me to remain in training after the recruit class graduated to introduce the oriented method to the entire department.

Figure 12-8. A search team getting ready to "hit Division 7."

The one-person oriented method of search and rescue is an alternative to the standard two-person search and rescue. In the standard method of search and rescue, a search crew enters the structure together and goes to the area to be searched. One or two firefighters go to the left, and the other firefighter(s) go to the right; they meet somewhere in the middle of the room. In other methods of search, a single crew member searches an individual room, and the crew meets somewhere in the hallway to move on to another area or to exit the structure.

Firefighters conducting a search and rescue operation inside a burning structure have two main concerns. The first is to deal with any possible victims. The second is self-preservation. The oriented method basically divides these two concerns: One firefighter has as his main concern or focus knowing how to get out at all times and ensuring the survival of his crew and himself. The other firefighter's only concern is to focus on the search.

The oriented method is not a panacea or the cure-all of search and rescues. Sometimes it is impracticable, such as in buildings that have large, open areas such as warehouses, factories, and restaurants or banquet halls. However, for residential single-family and multifamily structures (our most frequent structure fire type), the oriented method works extremely well. It also mirrors the incident management system: It prohibits the officer from participating in the actual search and emphasizes focus.

During tests in controlled situations (blacked-out mask with no smoke or fire), the following was found:

- The searches were conducted faster.
- Search teams got lost less frequently, if at all.
- The firefighter doing the search did a much more thorough job.
- The firefighter doing the search was much more at ease and confident.
- The same number of victims was found as in the standard method.

Controlled situations were used at first so that the searcher and the oriented man could be observed. Later, tests were done in actual fires. The results remained the same. The idea behind the oriented method of search and rescue is *focus*. The oriented man (usually the search sector officer) remains in a central area (usually the hallway) and focuses on his environment and his crew's safety, allowing searchers to focus on the search.

The most common argument against this method is that personnel who could be used for searching or other firefighting duties are wasted by keeping a firefighter in a hall or near a door. That just as many victims were found in a shorter time; crews got lost less frequently, if at all; and the searcher felt much more secure and confident knowing someone was watching out for him should dispel this argument.

Three things are necessary for a successful search and rescue using the oriented method:

- *There must be communication between the oriented man (the search officer) and the searchers.* The interaction not only lets the oriented man know where the searcher is at all times but also helps put the searcher at ease.
- *The direction of search must be determined (left- or right-handed).* When doing a left-handed search, the searcher uses the wall on his left as a way of knowing where he is in relation to the exit in that particular room. More will be said about this later.
- *The number of walls in the room must be determined.* Basically, most rooms in a residential structure have four or five walls. More will be said about this later.

When conducting a search using the oriented method, the two or three members enter as a team and remain in voice contact throughout the time the team is inside. The oriented man must concentrate on how he gets in and where he is in relation to his exit. It is his responsibility to find his way out at any time. Once the search begins, the oriented man remains at the door or in the hall, and the searchers go in and conduct the actual search. While the searcher is conducting the search, the oriented man can (and should) do the following:

- remain aware of the searcher(s) through communication;
- keep going over in his mind, How do we get out?
- stay alert to changing fire condtions (buildup of heat or smoke, signs of impending structural collapse, and so on);
- listen for fire and human noises; and
- coordinate with any other sectors on the same floor and with Command.

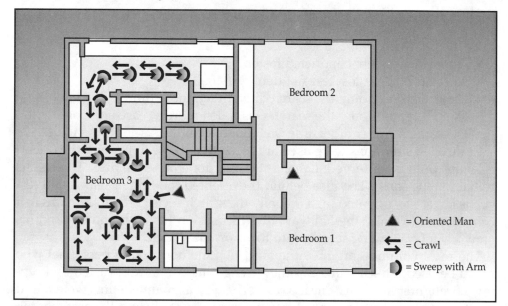

Figure 12-9. An oriented search of a bedroom on the second floor of a single-family home.

SCENARIO 1

Assume that the fire is on the first floor and that a two-person crew is going to search the second floor using the oriented method. They enter together and find the stairway together. The oriented man leads the searcher up the stairs and to the first room. The oriented man should work the left or the right side of the hall and not change. This makes for a more thorough search and also avoids confusion if a fast exit must be made. In this case, the oriented man goes to the right and leads the searcher to bedroom #3.

The searcher goes in, determines the number of walls, and tells the oriented man he will do a left-hand search. As the searcher goes in, he follows the left wall for about two steps, takes two side steps toward the center of the room, sweeps with his right hand, then takes two side steps back to his wall (the wall on his left). He does this until he reaches a door on wall #1. He feels the door for heat, then opens it, telling the oriented man what he is doing. He determines that it is a closet, sweeps the floor of the closet, quickly closes the door, and continues down the wall. He continues moving forward, taking side steps and sweeping and then returning to his wall until he covers the entire room and meets the oriented man at the door. At this point, the oriented man takes him to the next room to be searched.

When searching, remember that you are looking for something about the size of at least a one-year-old or a large doll. Children younger than this usually are not mobile and won't be found on the floor. Don't stop and feel every small thing you find.

This is a brief description of how the oriented method works. Now, we will get into some of the specifics that make this method work.

Direction of Search

One of the worst mistakes firefighters can make is to lose or forget their way out of a building. Unless you can see where you are going, it is a bad idea to walk or crawl into the center of a room. It's very hard for most people to travel in a straight line when they can't see where they are going. To turn exactly 180 degrees and return to the exact place you entered is almost impossible. To eliminate such problems, when entering a room, first decide if you are going to take the left or the right wall. Always keep that wall (sometimes referred to in this text as "your wall") to your left or right. Use that wall as your point of orientation for that room.

SCENARIO 2

Let's say you enter a room and go to the left. Your left hand or shoulder should be in contact with the wall on your left most of the time.

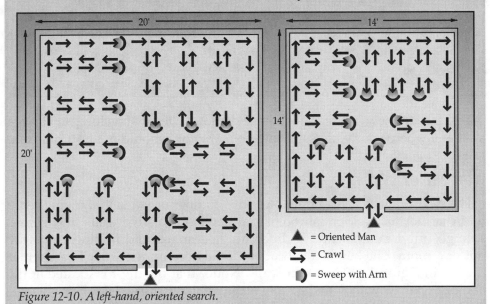

Figure 12-10. A left-hand, oriented search.

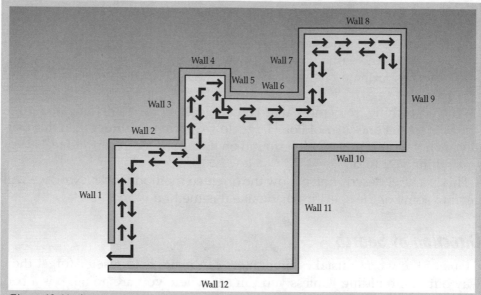

Figure 12-11. A very eccentric room with many walls and turns. By keeping track of "your wall," working your way back to the exit should be no problem should you have to escape in a hurry.

Crawl two steps down the wall, and then take two side steps toward the center of the room or to your right (see Figure 12-10). Sweep with your right hand, and then take two side steps back to your wall. Do this around the entire room until you get back to the door. By taking two steps forward, then two side steps, and sweeping, you can cover a 14-foot × 14-foot room and not miss a thing. If the room seems large to you after a few steps forward, you can increase your side steps to three or four. That way, you will be able to cover about a 20-foot × 20-foot room. If at any time you should have to get out in a hurry, all you would have to do is to go back to your wall, back out or turn around against the wall, and crawl forward, keeping your right hand or shoulder on the wall and following it back to the door. It doesn't matter whether you make 10 turns on the wall. If you count the walls and do not lose contact with your wall, you eventually will get back to the door (see Figure 12-11).

This is extremely exaggerated, but it demonstrates that as long as you are aware of where your wall is, finding your way out should not be a problem.

Number of Walls

The next thing to concentrate on is the number of walls in a room. Being told that normal residential rooms can have four or five walls may sound strange, but it can make a big difference in searches and finding your way out of a room. The key is the location of the door on the wall. The room at the left in Figure 12-12 has four walls. Notice that the door is located at the end of a wall. Now, look at the room at the right in Figure 12-12. This is a five-walled room. Again, notice the relationship of the door to its location on the

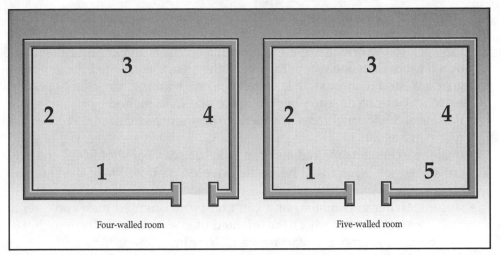

Figure 12-12. (Left) A room with four walls; (right) a room with five walls.

wall. The door is in the middle of the wall. (It doesn't have to be right in the center of the wall to be a five-walled room.) If you believe that most rooms (especially bedrooms) have four walls and you were working on the fourth wall in the room on the right in Figure 12-12 (the five-walled room), you would expect the door to be on the fourth wall. But, all you would find on that wall would be wall and corner. At exactly what point would you begin to panic? However, if prior to "working" the room, you had determined that this was a five-walled room, you would have expected the door to be on the fifth, not the fourth, wall.

If the searcher tells the oriented man that he is on Wall 5, the oriented man should know exactly where the searcher is and exactly where to head if the searcher gets in trouble or finds a victim.

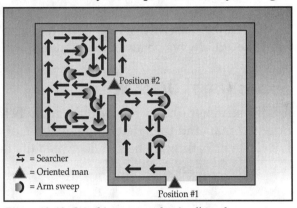

Figure 12-13. Searching a room that is off another room.

OTHER FINE POINTS

- The oriented search works best if the search crew consists of two or three members (the officer or the oriented man and one or two searchers). Communication can become difficult and confusing to the oriented fire-fighter when there are more than two searchers.
- The oriented man and searcher(s) must not switch jobs in the middle of a search. Doing so would make finding the way out more difficult. One

person, and only one person, should be responsible for knowing how to get out.

- If the searcher finds a victim who can be moved, he should drag the victim back to the oriented man, who will lead the searcher and the victim out. If more than one victim may be in the area, the oriented man should go back and continue the search, either teaming up with another searcher or team or searching using the one-man mehod.
- If the searcher cannot move the victim, the oriented man should enter the room and assist.
- If the searcher gets injured and can talk, the searcher should tell the oriented man on which wall he is; the oriented man will take the shortest route to him.
- If the searcher gets injured and can't talk (the oriented man can't get a response from the searcher), the oriented man should go to the searcher's last known location and start to search for him.
- If the oriented man gets hurt and can talk, the searcher should backtrack to the door or hall where the oriented man is; the oriented man will tell the searcher how to get out if he doesn't know the way.
- If the oriented man gets hurt and can't talk, the searcher should backtrack to the oriented man's location and
    —if he knows the way out, take the oriented man out.
    —if he does not know the way out, take the oriented man back into the room being searched and go to the last known window and summon help.

Note: The chances of the oriented man's getting hurt are slim. He is usually in the hall or doorway, which usually doesn't collapse. Also, the oriented man should always be aware of changing fire conditions.

Rooms Off of Other Rooms

If the searcher finds a room off the room being searched, he should tell the oriented man and then determine the type of room he has encountered. If it is a closet or bathroom, he should sweep the floor area quickly and then con-

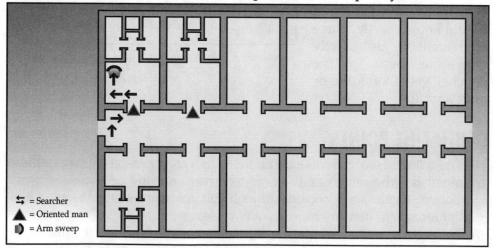

= Searcher
= Oriented man
= Arm sweep

Figure 12-14. Searching in large apartment buildings.

tinue to search the original room. If it is another bedroom, the oriented man should go to the new room and stay at the doorway while the searcher searches the new room. When that room has been searched, the oriented man should go back to the original door or hall and the searcher should continue to search the first (original) room (See Figure 12-13).

Large Apartment Buildings

The oriented man should count the number of doors as he passes them. If two teams are on the same floor, one team can take one side of the hall and the other the other side. If only one team is on the floor, members should always try to work the same side of the hall. This will avoid confusion if a fast exit must be made (see Figure 12-14).

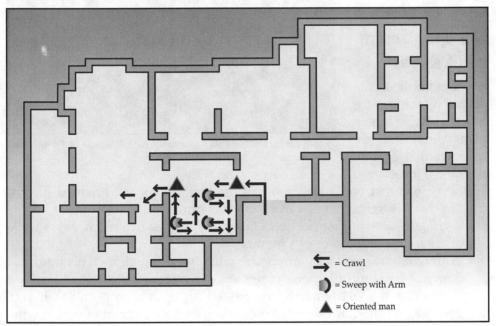

= Crawl

= Sweep with Arm

= Oriented man

Figure 12-15. Searching a ranch house demands concentration and imagination.

Ranch Houses

The number of rooms off other rooms in ranch houses and on the first floor of a two-story home can create problems. The searcher should do a left- or right-hand search until coming to another room, constantly informing the oriented man of his location. When the searcher comes to another room, he should determine the type of room it is. If it's a bathroom, closet, or the like, he should sweep it quickly and then move on. If it is a living or sleeping room (the time of day, of course, would dictate priorities), the oriented man should go to that room's entrance and the searcher should enter and search the room. The searcher should continue to the next room while the oriented man remains at the last room searched. When the next room is found, the oriented man will move up again. He must keep in mind how to get out if necessary. Coordination between two or more search crews will help this situation.

As you can see, this type of search calls for much concentration and imagination on the part of the oriented man (see Figure 12-15).

Figure 12-16. During a secondary search, beds should be turned over and furniture moved.

One-Man Search and Rescues

One of the most dangerous operations for a firefighter is a one-man search and rescue. Avoid it if at all possible. However, if you must conduct one, using some of the methods of the oriented search, such as those listed below, can make the task safer.

- Use a left- or right-hand search. Unless you can see in a room, never walk or crawl into the center of a room.
- Count walls. This is very important when working alone. This information may be your only point of reference in relation to your exit.
- Remember the last window you passed (or door, if applicable). If you need to get out, don't advance unless you can see in front of you. A window or door may not be close enough—if there is one at all. Back out the way you came in.
- Remember what you are passing, such as furniture. If you are backing out and don't run into it, you may have done something wrong.

The oriented method is a new technique; many variations and methods can be used. As noted, it is an alternative method and not the last word in search and rescue. However, it has proved to be a good method so far and deserves a try. The choice is yours.

Endnotes

1. The responsibilities of Backup are discussed in detail in Chapter 13.

2. Reading trade journals and viewing training-oriented videotapes can help you to prepare for your next incident. These case studies can prove invaluable as training and "nonparticipatory experiential" learning tools.

3. *Fire Protection Handbook*, 17th edition, second printing. 1991. National Fire Protection Association (NFPA) Publications.

4. "Survive the Search and Rescue," author unknown, *Fire Service Today*, May 1983.

REVIEW QUESTIONS

1. What is the mission of Search?

2. What are the three responsibilities of the search officer?

3. What are the two basic search strategies?

4. Define "primary search."

5. Define "secondary search."

6. What are three assumptions Command must make concerning the search?

7. What assumptions must the search officer make?

8. What four indicators can be used to help determine the best areas for beginning your search?

9. Should the search team members have a protective line with them? ___ Yes ___ No

10. Should every fire victim be taken outside for revival? ___ Yes ___ No

11. What is the main difference between more standard methods of search and the oriented method?

12. What three things must be done when conducting a search using the oriented method?

13. What is the key factor associated with the number of walls in a room?

14. Why should the oriented man and the searcher *not* switch positions during a search?

15. Why is it difficult for the oriented man to have more than two searchers with him?

DISCUSSION QUESTIONS

1. Discuss the correlation between savable victims and response time. What is the average response time in your jurisdiction?

2. Divide the class into two groups. Have one group justify a search crew's taking a line with them. The other group should justify why the search team should not take a hoseline with them. Appoint spokespersons and present your arguments.

3. Discuss and compare your department rules for removing fire fatality victims with the author's suggestions.

4. Compare and contrast your department's "preferred" or standard method.

13

The Mission of Backup

TOPICS

- The mission of Backup
- The relationship between Command and Backup
- The responsibilities of Backup
- When the backup line should be off an alternate water supply
- Where and why the backup line should enter the building
- Where to position the backup line
- When Backup can be used for other assignments
- When one backup line may not be enough
- The difference between rapid intervention and backup sectors

KEY CONCEPTS

✔ Alternate water source
✔ Backup
✔ Backup engine
✔ Dry line
✔ Opposing lines
✔ Working crews

The mission of Backup is to protect the interior working crews (attack, search, and other interior firefighters) by pulling and strategically placing protective lines between them and obvious or potential areas of fire spread and to ensure egress if retreat is necessary.

The last thing Command wants the backup crew to do is to seek out and extinguish fire. The backup sector provided the impetus that led our department to create a mission statement. Crews kept telling us about fires at which the backup crew was used as a second attack line. Stupid us on the outside! We thought that the officer and crew assigned to backup were inside looking out for the interior crews.

Backup's only purpose is to watch out for the safety of interior working crews. He cannot effectively do this if he is looking for and extinguishing fire.

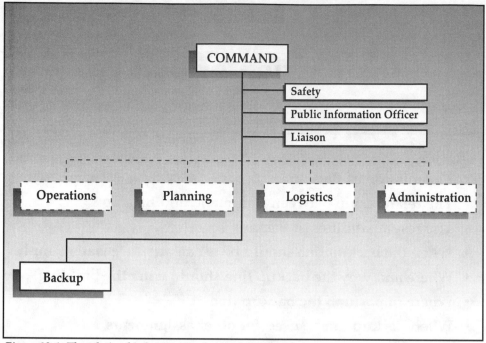

Figure 13-1. *The relationship between Backup and Command.*

SCENARIO 1

The crew on line #2 is the backup crew. Backup has stretched his line into another room to put out a small fire. If Backup's attention is on putting out that small fire, can he be focused on the safety of the interior companies? The answer is, No.

THE RELATIONSHIP BETWEEN COMMAND AND BACKUP

I like to think of Backup as *interior safety*. The backup team's two primary concerns should be ongoing monitoring of interior conditions and knowing the locations of all companies operating within the structure. Command should assign a backup sector anytime a crew takes inside a line larger than a booster line (one inch). Procedure should dictate that initially no line smaller than 1½ inches be pulled to attack a fire inside a structure. If Attack feels that a 1½-inch line is too large for the size of the fire, he can call for a booster line and keep the 1½-inch inside as a backup. That means that a backup sector should be assigned at every working fire at which a line is pulled. After Command has assigned a backup sector, he can "relax" because he knows that inside is a crew that is focusing solely on protecting all the interior crews.

Command should have a good idea of where all the companies he has assigned are operating. Attack should be on the fire floor (or on a specific floor) with the intention of moving up to the next floor as soon as the fire on

Figure 13-2. Backup cannot focus on the safety of interior crews if he is putting out a fire.

the floor he is on has been knocked down. Search should be in the area where victims *should* be and where their chance for survival is the greatest. The backup line should be positioned in a location well suited for covering interior crews. That may be a specific floor or in between floors. Command and Backup must trust each other. Backup should be a seasoned officer or firefighter Command believes will be able to identify potential trouble signs and move away from his crew and the line for short periods of time to check on other crews. The backup officer must trust that Command will notify him of any changing conditions observable from the outside.

THE RESPONSIBILITY OF THE BACKUP OFFICER

Backup has three responsibilities:

- *The safety of all interior crews.* Each sector officer's primary responsibility is the safety of his crew. The officer assigned as Backup shares that responsibility. Just as the relationship between Command and Backup must be based on trust, so too must the relationship between the officer assigned to backup and his crew. As noted, the backup officer may be remote from his crew for periods of time (the reason for this will become evident later in this chapter). Firefighters assigned to backup, therefore, must know that they must stay put and hold their ground unless their officer directs them to do otherwise.

- *Position and maintain a working backup line.* The backup officer's first tasks are to determine a suitable location for the backup line that will protect interior crews and to locate a means of egress. Next, he takes his crew to the location of the backup line and gives them detailed instructions. He then leaves the area of the backup line (he may take another crew member with him if one is available) and moves from work area to work area within the structure to ensure that other interior crews are safe and their means of egress is intact.

• *To keep Command informed.* Backup must tell Command when interior conditions change. More than anyone else, Backup must monitor the conditions within the structure and report to Command any changes in heat (a buildup or reduction) or smoke (color, intensity, or direction of movement). Comparisons between interior and exterior conditions are necessary components in the on-scene size-up and in prioritizing the assignments on Command's to-do list.

If the backup sector is in place and should find fire, Backup can do one of two things:

• If the fire is small or nonthreatening, he can inform the attack sector by radio or in person of the fire and its location.

• If the fire is of significant size or is in an area that has the potential to threaten working crews or considerable property, Backup should inform Command of the fire's size and scope.

On receipt of this information, Command will do one of the following:

• send in a separate attack company to work the fire;

• request Attack to move to the fire area as soon as he can;

• inform Backup that another company is being sent to Backup's location and that, when it arrives, Backup will have a new division/sector/attack designation; or

• tell Backup to knock down the fire quickly and that someone will be sent there shortly to overhaul the area. If this is the case, a member of the backup sector should stay in the area in which the backup line is located and attempt to remain focused on the safety of the interior crews for that brief period. The backup officer must be aware of time and avoid focusing on the fire longer than is necessary. If possible, doors should be closed to hold the fire in that specific area until water for attack becomes available. This last option should be used only as a last resort.

ALTERNATE WATER SOURCES

In rural America, it's sometimes difficult to supply the attack line from a continuous water supply. Booster tanks may be the only option for an alternate water source. Extra care must be taken to keep working crews out of harm's way. In urban settings, a supply line taken from a separate hydrant should suffice as an alternate water supply. If water is available and the fire's intensity or its hold on the structure is a concern, I have no problem with having the engine assigned to backup take its own hydrant and lay in. That way, if the engine supplying the attack company fails or if its water supply is disrupted, water will still be available to protect interior crews until they are able to back out of the structure.

I order the backup crew to secure its own water supply (preferably from its own hydrant) when the following conditions are present:

• The fire is at or near the flashover stage and I have committed to an offensive strategy.

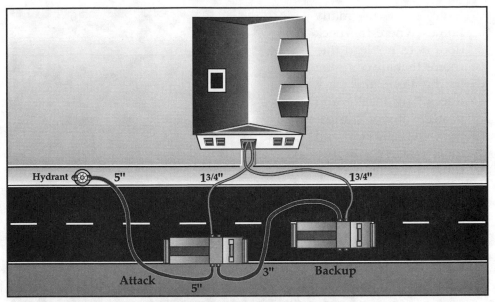

Figure 13-3. Backup is securing water from the attack engine's hydrant.

- Multiple fires are on multiple levels of the structure.
- The fire has entered and is attacking the structure itself. If the structure is of balloon-frame construction, I make every effort to have the backup crew lay in its own line.
- The fire is in a large structure and multiple backup lines are in place.
- There is a working fire in a commercial occupancy.

If the fire is a one-room fire or one I feel can be handled by one attack line and I believe there is no unusual risk to personnel, I take a line from the attack engine to supply the backup engine (Figure 13-3). If the attack engine or its source of water were to fail, I would still have 500 gallons of water in the booster tank of the backup engine that could be used to back out crews. If used correctly, 500 gallons is a lot of water. *If there is any doubt about the safety of crews, an alternate water supply should be established for the backup line(s).*

FROM WHERE SHOULD BACKUP ENTER THE BUILDING?

As a rule of thumb, the backup line should follow the attack line into the structure. If the attack line was taken through the front door, the backup line should also be taken in through the front door for the following reasons:

- *It decreases the possibility of opposing lines.* If the backup line follows the attack line in, crew members will be less likely to be hit by a hose stream.
- *Backup's purpose is to protect interior crews and provide a way out of the structure.* Normally, when crews back out of a structure, they keep their line with them or follow the line out. If the backup crew takes a different way into the structure than the attack crew, the backup crew may not even know which way the attack crew came in, let alone where they are. I

have been in many homes where the back door does not lead into rooms that are common to the front and back doors. To locate and protect interior companies, the quickest and most efficient entrance route for the backup sector and its line is the same entrance route used for the attack lines.

WHERE TO POSITION THE BACKUP LINE

Unless it is necessary to protect firefighters who are in imminent danger or the distance between the working crews (attack, search, extension, salvage, and so on) and the initial placement of the backup line has in-

Figure 13-4. Backup should take the same way into a structure that Attack takes.

creased to the point that it would be ineffective, I do not advocate moving the backup line once it has been placed. If properly located initially, it will rarely have to be moved during the typical single-family residential structure fire.

The proper location of the backup line and its crew depends on the following:
- *The number of floors in the structure.* If the structure has only one living floor and the fire is on that floor, the line should be placed approximately two-thirds the distance between the point of entry and the location of the attack crew.

Shown in Figure 13-5 is a single-family, ranch-style home that has a fire in a rear bedroom. The proper location of the backup line is somewhere near the entrance to the hallway. This leaves the attack crew plenty of room to work; Search (if assigned) an avenue for moving in and out; and Backup room to move among the backup line, the working sectors, and the likely means of egress (the front door). Once positioned, the line should remain stationary.

If the structure has two floors, the backup line should be positioned according to the locations of the fire and the working crews. These situations are covered later in this chapter under "Location of the Fire."

The officer assigned to backup should give his crew instructions that include the location to which he is going ("I'm going to follow the attack line up to Attack. I'll be right back") and reminders to watch for changing conditions ("Keep an eye on the basement stairs; it may be below us"). I advise

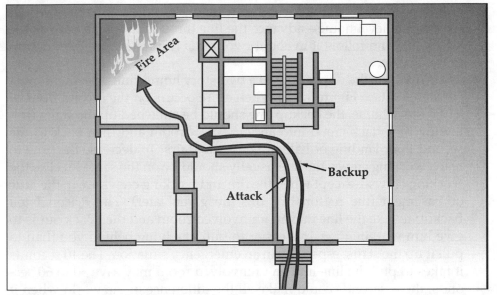

Figure 13-5. Fire on the first floor of a one-family or two-story home.

using a "count" method in extreme conditions ("I'll be back in less than 45 seconds"). If the crew doesn't hear from the officer in about 45 seconds (counts), someone will go to check on the officer or advance the line. (The officer should be following the attack line up to the attack crew.) Under normal conditions, I tell the backup crew to just "camp out" at the location their officer has chosen for them and not to leave that area unless directed to do otherwise or if problems should arise.

   • *The location of the fire (as it relates to floor level).* If the fire is on the first floor

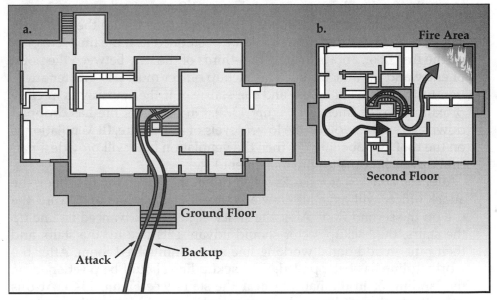

Figure 13-6. (a) The lines were advanced into the house and up to the stairway for this fire on the second floor of a two-story house. (b) Placement of the lines on the second floor.

of a two-story home, backup crews should place the line as they would for a single-level home—advance the line through the same path as the attack line and follow it in about two-thirds from the point of entry, up to the nozzle.

If the fire is on the upper floor of a two-story home, line placement must be changed (see Figure 13-6). When a fire occurs on the second floor of a two-story home, the backup line should be stretched up the stairs (following the attack crew) into an *uninvolved* room and then back to the second-floor landing or hallway near the stairs. To decrease the time to reach working crews, the line usually should be on the same level as the working crews—except when the fire and working crews are in the attic or basement (the reasons for this are given later). The reason I tell Backup to take the line into an uninvolved room and then back out is to give him working line. It is easier to pull line lying behind you than to pull it up the stairs, especially in an emergency situation. The 10 seconds it takes to pull the line into an uninvolved room may save 30 to 60 seconds; those seconds could make all the difference in the world when it comes to the safety of working crews. Figure 13-6 shows the correct placement of a backup line for a fire on the second floor of a two-story dwelling. Note that the line was taken in and up the same way as the attack line. If at the top of the stairs the attack line goes to the left, the backup line should go to the right to find a room to pull some "working line."

- *If the fire is on more than one floor.* When there are multiple fires involving more than one floor, the job of the backup sector becomes much more demanding. As stated in Chapter 11, if fires are on different levels of a house, the rule of thumb is to start at the lowest level and work your way up. If the fire involves the first and second floors of a two-story house, the attack sector usually will head first toward the fire on the first floor. The backup line at that time should be positioned as if the fire were only on the first floor, approximately two-thirds of the way between the point of entry and the attack crew. The backup officer must pay greater attention to the means of egress and the stairway if the stairway is in close proximity to the lines (see Figure 13-7). On occasion, fire has dropped downstairs and involved the lower levels of a structure. (If Ventilation is on the roof and opening up, then the ventilation hole will provide a natural upward means for fire movement.)

After the attack sector has knocked down the fire on the first floor, the attack officer will have his crew advance the attack line up toward the fire on the second floor. After the attack sector has advanced its line up the stairs, the backup sector should advance its line up the stairs and then pull an additional working line in an uninvolved room. After the working line has been pulled, the backup line should be positioned on the landing or in the hallway near the stairs (see Figure 13-7). At this time, the backup officer and a crew member should ensure the safety of the attack crew and then quickly move back downstairs to ensure that

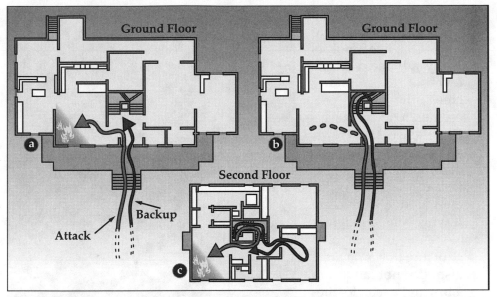

Figure 13-7. (a) Units are attacking the fire on the first floor first with the placement of a backup line. (b) The lines were moved up the stairs after the fire on the first floor was knocked down. (c) The backup line is on the landing near the stairs.

they and the attack crew are not going to be cut off by any flare-ups or hidden fire downstairs. The backup officer should move back and forth between these potential problem areas until Command declares the fire under control and the attack officer has given his benchmark[1] "the fire is knocked down."

If the backup officer is concerned about the intensity of the fire down-stairs, he may move the backup line back down the stairs to avoid being cut off upstairs. If he moves the backup line downstairs, he must ensure that there is adequate line to quickly advance back up the stairs if it becomes necessary.

- *Fire in the basement.* The backup line should be positioned near the top of the basement stairs (see Figure 13-9), and an additional working line must be pulled so advancement is not hampered. There is little likelihood of opposing streams in a basement; there usually is only one way in or out of a basement. Basements are wide open, and the attack crew will advance directly on the area of the fire. If the intensity of the fire or smoke in the basement gives the backup officer concern, the line should be taken downstairs. Keep in mind that basement stairs are not as wide as other stairways and have smaller railings to steady firefighters if they should lose their balance while ascending or descending. The backup line should be moved to the side of the stairs so firefighters don't trip over it if they must make a quick exit.
- *Fire in attic areas.* If the fire is in the attic, the backup line should be positioned at the bottom of the attic stairs (see Figure 13-10). Additional line must be pulled onto the second floor so that advancement of the line will

not be hampered.

In attic and basement fires, if the line is not actually advanced to the level of the working crews, the backup officer must rotate among his line, the working crews, and the means of egress, or he must assign that task to members of his crew. I don't advocate leaving one firefighter alone in an area of a structure that is on fire. I also do not advocate allowing a single firefighter to roam a structure on fire. This creates a problem. If the backup sector is a four-person crew, two members can be left on the backup line while the officer and a crew member go among working sectors

Figure 13-8. The backup sector is placing its line as units are attacking the fire on the first floor.

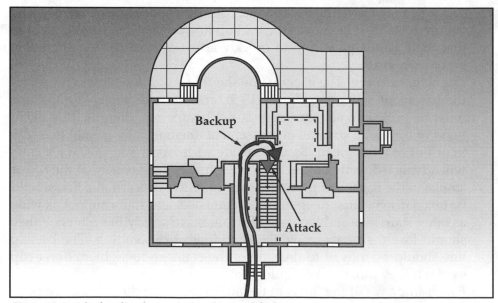

Figure 13-9. A backup line is stretched to the top of the basement stairs.

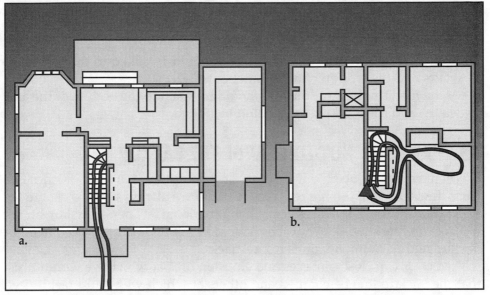

Figure 13-10. (a) The lines for this attic fire are going in at the first-floor level. (b) The attack lines going to the second floor. Note that the attack lines move up to the attic stairs while the backup line pulls extra line and remains on the second-floor landing.

of the fire. However, if the backup sector is a three-person crew, a decision must be made as to who will be alone. One possibility is to assign another firefighter to backup as soon as one is available. Until then, the backup officer should remain within view of his crew (on the line). Another possibility may be to leave the single member on the backup line with a portable radio. He has a line for protection and should be within 10 to 20 feet of a working sector. That way, the officer will have a member with him while he goes among sectors. The decisions relative to the number of personnel on lines and who can be left alone should be up to the department chief and be incorporated in a written procedure.

REVIEW

- If the fire is on the first floor of a single-story home or the first floor of a two-story home, the backup line should be advanced approximately two-thirds of the way between the point of entry into the structure and the attack crew.
- If the fire is located on the second floor of a two-story home, the line should be taken up the stairs into an uninvolved room and then positioned at or near the stairs.
- If the fire is on more than one floor, the line should be taken up to the location of the attack crew. When the attack crew initially enters the first floor, the lines should be positioned approximately two-thirds of the way between the point of entry and the attack crew. When the attack crew has

knocked down the fire on Division 1, the line should follow the attack crew up the stairs into an uninvolved room and be positioned at or near the stairs.

- If the fire is in the basement, the line should be taken to the top of the stairs and the working line should be pulled.
- If the fire is in the attic, the line should be taken to the bottom of the attic stairs, and the working line should be pulled.

TO CHARGE OR NOT TO CHARGE THE BACKUP LINE

When we introduced the concept of backup lines in the Toledo Fire Department, one of the big concerns was, Do we always need to charge the backup line? To me, the answer is, "It is situational." (The Safety Bureau felt the backup line should always be charged.) I felt it was up to Command or Backup and recommendations from Attack.

Most of us can close our eyes and envision fires at which we would insist that the backup line be charged and others at which we would feel it need not be charged. Command is ultimately responsible for the outcome of the incident, including the safety of the working crews. If Command does not assign a backup sector (if personnel are available) when fire conditions dictate that a backup line be pulled and charged, Command should be held responsible if a firefighter is injured because a backup line is not in place.

If the attack sector gets inside and makes a quick knockdown and the extent of the fire is small, I have no problem with Attack's telling Command that a charged backup line is not needed.

Attack to Command: We have the fire knocked down. Hold up on charging the backup line.

Command: Okay.

Command to Backup: Get the line in place, but we'll hold up charging the line unless you tell us to.

Backup: Okay.

In most single-family house fires, we allow the backup line to be positioned before it is charged. It is much easier to pull a dry line. Once the line is in place, the backup officer can make one of two transmissions:

Backup to Engine 13: Charge the backup line,

OR

Backup to Command: The backup line is in place, but I'm going to hold up on charging the line. It looks like they got it.

Command can always override that decision and order that the line be charged. If that is the case, the line will be charged and any concerns will be discussed after the fire.

SWITCHING THE BACKUP LINE TO AN ATTACK LINE

As noted, our department developed and implemented mission statements in response to companies' accounts of how crews assigned as backup aggressively sought out and attacked fire.

If the backup sector does what it is supposed to do, it might get to open the bail of the nozzle once in a thousand assignments, if at all. Normally, the backup sector is a safety valve. In reality, backup can be a boring assignment. Most fires are small, and usually actions to protect interior crews are not needed.

About the time we were learning it was "customary" for the backup sector to be looking for fire to put out, we had a rash of fires involving attic areas. At several of these fires, the attack sector called for additional lines because of the difficulty in making the attic. We learned that at several of these fires, the backup line was used for this additional firefighting power. That could have been a fatal mistake.

If the backup sector leaves its position to work as an additional attack line in the attic stairway area, who would be looking out for the crews? No one! At one fire, the attic flashed over, and several firefighters came out of the attic area with their bunker coats on fire. After tempers (and a few individuals— literally) cooled down, we looked at this problem. If you have an attic "rolling" to the point where attack needs another line inside to help the crew advance, I do not believe that is the time to pull out the backup sector and convert it into an attack sector.

Now, I'm not naive, and it hasn't been that long since I was a line officer. I realize the temptation that is present when you're standing 10 feet from a company yelling for more water (another line) and you have a charged line in your hands. But the safety of your crew and the working crews inside must take precedence over the temptation to open that bail.

We have compromised with our officers to an extent. In the instance stated above (Attack is yelling for another line, and the backup crew is standing there with a charged line in their hands), a little communication, common sense, and coordination should solve all the concerns here. If Attack asks for another line, we allow the backup sector to assist Attack, *once another backup company has been put in place.* It is impractical to have a crew "pass" another crew inside a building. However, if a fire is burning with such intensity that additional hose streams are needed to extinguish it, it is unwise to remove backup or even delay or postpone its operation. Common sense must prevail over aggressiveness to ensure the safety of interior companies.

USING BACKUP FOR SUBSEQUENT ASSIGNMENTS

The backup sector should not be removed until someone has checked for fire exten-

sion. Until you're sure the fire hasn't spread to other undetected areeas, leave the backup sector in place. No matter what the assignment, the backup sector should not be terminated until the following has occurred:

- all visible flame has been extinguished;
- Attack has given the benchmark "The fire is knocked down";
- all areas have been checked for extension;
- Command has observed that no smoke is "forcibly" leaving the structure, that the color of the smoke has lightened; and, most importantly,
- Command has given the benchmark "Under control."

Once Command believes the fire is under control, he can consider using the backup line for other assignments. Typically, the backup crew will be reassigned to assist Attack with overhaul.

PROTECTING OTHER SECTORS

The majority of this chapter has dealt with the relationship between the backup and attack sectors. Other interior sectors can and should be working inside the structure. The search sector comes to mind first. As stated in Chapter 12, I do not believe the search sector should take a hoseline with it because if Search pulls a line, the number of available searchers is decreased (someone has to stretch and hold the line; a line lying in the hall when you're in the bedroom does little good), and the search crew is forced to divide its focus between victims and nozzles.

This option is unnecessary, moreover. In the normal house fire, backup can position itself to protect the attack and search sectors even if they are operating in different areas of the structure. Most of the same rules that apply to protecting Attack apply also to protecting two different sectors:

- If the attack and search sectors (working crews) are working on the first floor in the same basic area, the backup line should be about two-thirds between the point of entry into the house and the working crew.
- If the attack and search sectors (working crews) are working on the second floor, the backup line should be pulled up the stairs and the working line pulled into a previously searched room or down a hallway. Be careful not to interfere with the search sector. If the second floor is small, the nozzle can be taken up the stairs and a firefighter can be left down at the bottom of the stairs to feed line up if it is needed. Common sense should dictate.
- If the attack and the search sectors are operating on different floors, the backup line should be positioned as if fires were on both floors of the structure—up the stairs with an additional line with which to advance. Since the attack crew has a line and the search crew does not, if there is a need to choose which sector to stick closer to, the edge should be given to the search team, which has no line at all and needs protection the most. The location of the fire and Search's proximity to a means of egress will help determine where to place the backup line.

Other interior sectors that may need protecting are extension, overhaul, salvage, and in some instances, ventilation. All are part of the working crews.

IS ONE BACKUP LINE ENOUGH?

In some situations, the intensity of the fire and its hold on the structure may dictate that multiple attack lines, multiple search crews, and subsequently, multiple backup lines be employed. As Command, I have established two separate backup sectors at several fires. Just as there can be a Division 1 and Division 2 Attack, there also can be a Division 1 and a Division 2 Backup. If the size of a specific floor is large or cut up with fire inside the structure (inside walls and running floor or ceiling joists), a Sector B and a Sector C Backup may be assigned.

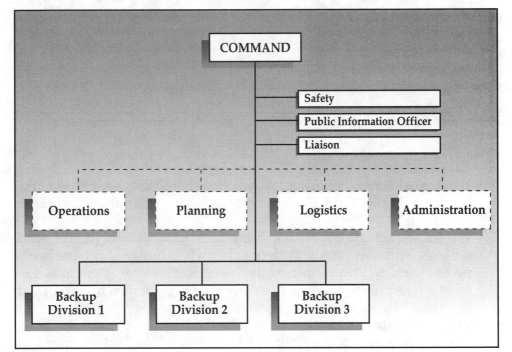

Figure 13-11. Backup sectors assigned to protect crews on specific floors (divisions).

When multiple backup lines are used, more than one sector officer may be assigned to manage each backup sector, or the backup sector may be under the direction of the same sector (branch) officer. The approach to be used would depend on the following factors:

• *The span of control.* No officer—especially a backup officer, because he may have to leave his crew unattended for short periods of time—should have more than five subordinates. However, if a five- or six-person crew is assigned and the officer designated as backup can control the activities of two separate backup crews working in different areas of the structure,

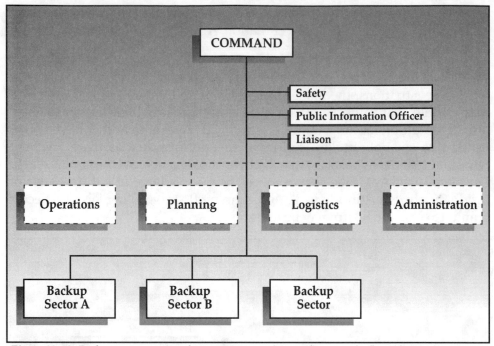

Figure 13-12. Backup sectors assigned to protect crews in specific areas of a floor (division).

one officer may be sufficient to control both lines.

- *The distance between the individual backup lines.* The distance between the two (or more) backup lines may make individual control impractical or impossible.
- *The tasks being covered.* Attack companies have their own line (and defense). This does not nullify the need for a backup company. One backup line can cover two or more attack sectors. However, if multiple sectors that normally do not use hoselines are assigned, then multiple (and individual) backup lines may be warranted.

If two or more backup sectors are required, it may be advisable to designate backup as a branch. As Figure 13-13 illustrates, the backup branch is directly under the control of one officer, the branch officer, and each specific backup sector also has an officer assigned to it.

The backup branch officer may need to move from working area to working area to ensure that his sector officers are providing the best coverage possible to their specific working crews.

ASSIGNING BACKUP FIRST[2]

To put it simply, a backup should be assigned every time you go inside a structure *to fight a fire*. No one should argue that using a backup sector doesn't make sense. Firefighters perform their tasks better and quicker and feel better while doing them if they know someone is there looking out for them. There are personnel concerns; but more than anything else, the fire condi-

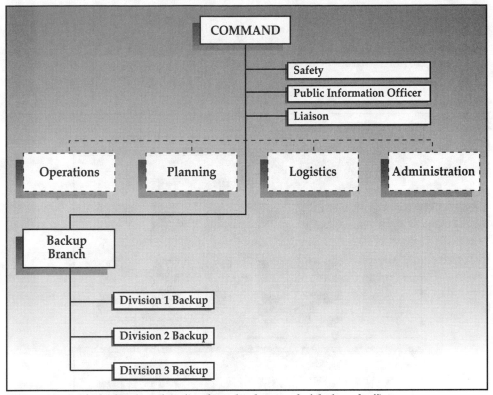

Figure 13-13. The backup branch is directly under the control of the branch officer.

tions will dictate where backup falls on Command's to-do list. I recommend that a backup sector also be assigned anytime Command orders an attack sector.

Command determines when backup comes to the top of his to-do list. Sometimes the complexity of the situation will push aside the thought of assigning a backup. In those situations, it may be even more vital that a backup be assigned immediately. I have been to fires where backup was the first assignment. Keep in mind that the backup sector is basically a stationary sector. If the situation calls for an offensive attack strategy with numerous or unknown rescues, it may be prudent to get a backup sector established from the onset of the operation. The line might be positioned in a stairwell and be expected to remain in that stairway for the duration of the incident until Command gives the "under control" benchmark.

SCENARIO 1

You are the first officer to arrive at a structure fire in a large four-story apartment building in the central district of the city. There are 16 living units (four on each floor and the ground level). You know the structure well. It seems as if you make an EMS run to this address every day. The halls are narrow. The walls are lath and plaster. Only one stairway provides access to the upper floors. Fire escapes are in the rear (off the alley), but they are of wood and their stability is questionable.

On arrival, heavy smoke is coming from Division 4 Sector A. A flicker of fire is on Side 2 of the structure. It is 0145 hours on a Sunday morning. A full assignment (three engines, a truck, a heavy squad, and a chief) is responding.

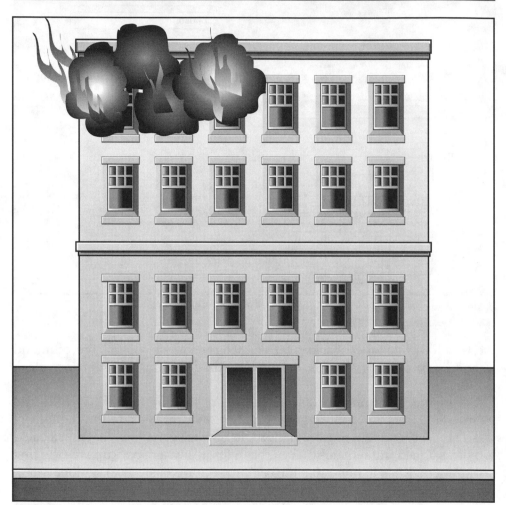

Figure 13-14. Fire in an apartment building where Command may choose to assign the first sector as backup.

In the above example, the first officer on the scene faces an obvious working fire. He knows the structure, the occupants, and the fire potential. The strategy at this fire would be offensive. (To go defensive at this fire would only push the fire around inside the structure and afford little opportunity to rescue any occupants.) It would be wise to make the backup sector the first assignment at this fire.

Figure 13-15. The backup crew is protecting the interior stairs in this apartment fire.

The radio transmission would be similar to the following:

Engine 5 is at 1945 Vermont. We have a working fire in an occupied apartment building. Engine 5 is Vermont Command.

Command to Engine 5: You're Backup!

At that time, Command would tell the crew to take a line inside and protect the stairs. (In doing so, hopefully the backup crew would pull a charged line into the structure and find the fire apartment. If the door to the apartment is not closed, they should attempt to close it. The chance of finding a viable victim in the involved apartment is slim, and the effort must be weighed against protecting savable victims. From that point on, the backup sector should do everything to prevent the fire from spreading into the hall and the open stairway for the following reasons:

- Occupants attempting to leave their apartments will have an easier time descending the stairs (if that is their chosen exit route[3]) and exiting.
- Crews assigned to search will have an eaiser time making the upper floors (and working there) if they know a crew is preventing the fire from entering the stairs.
- Once an attack sector has been assigned, it can open the door to the fire-

involved apartment and start to work on controlling the fire.

- At the time the working sectors enter, they already have a backup crew assigned and working.

The order in which sectors are placed on Command's to-do list at this fire is indeed out of the norm. However, a logical approach was given to an extreme problem. In this case, the positioning and application of one hoseline with the objective of holding a fire to a specific area while keeping the stairs operable allowed occupants to escape under their own power and provided a safe means of access for firefighters to trapped or frightened occupants who chose not to leave.

For example, let's say

Figure 13-16. This backup crew is pulling in a 2½-inch line to back up Attack.

that Command assigned the first crew as attack. The attack sector would have taken the line into the structure by way of the front door and up into the involved apartment. As the crew advanced on the fire, the first problem it would have encountered was that the door to the fire apartment could not be closed because of the hoseline. Consequently, smoke, heat, and fire would have passed into the hallway and eventually banked down the stairs. Remember, it's not *always prudent* to put out fire as soon as possible.

WHEN TO ASSIGN BACKUP

Assigning backup is a situational decision. Normally, backup is assigned after the attack sectors, and perhaps even search sectors, have been assigned and have begun working. However, there may be times when Command may want to assign a backup crew immediately on arrival. The crew could be used to do the following:

- protect the means of egress and access if the location and involvement of fire warrants, as in the above scenario;

- provide a protective line for search crews if the first priority (item on the to-do list) is search; and
- ensure a safe and continuous means of egress for interior crews if the fire involves more than one level or area and it could affect the outcome.

RAPID INTERVENTION TEAM (RIT)

The rapid intervention team (RIT), discussed in Chapter 18, is a relatively new concept in the fire service. The National Fire Protection Association (NFPA) recommends these units be used in some aspect at all structure fires. Backup must not be confused with rapid intervention. They are separate and distinct functional sectors and must stand alone.

Endnotes

1. Benchmarks are specific announcements given by sector officers. They are covered in Chapter 19.

2. Backup sectors may be assigned at incidents other than structure fires. These incidents are explained within their specific areas. This section covers only structure fires.

3. Civilians normally attempt to leave a building from the same location they entered. Unconscious occupants are most often found within the area leading to the exit they selected.

REVIEW QUESTIONS

1. What is the mission of Backup?

2. What are the responsibilities of the backup officer?

3. Give two examples of situations where Command may want an alternate water source for a backup line.

4. Where should Backup enter the building?

5. The proper location of the backup line and its crew depends on what factors?

6. Where should the backup line be positioned in a ranch house?

7. Where should a backup line be positioned when the fire is on the first floor of a two-story house?

8. Where should a backup line be positioned when the fire is on the second floor of a two-story house?

9. Where should a backup line be positioned when the fire is in the basement?

10. Where should a backup line be positioned when the fire is in the attic?

DISCUSSION QUESTIONS

1. Discuss the positioning of the backup line as compared with the attack line. Do you agree with the author's reasons in support of this position? Why?

2. The author gave one example in which the backup sector may extinguish fire for reasons other than protecting crews. Do you agree with his reason? Do you believe this position is too strict or not strict enough?

3. The author believes no officer or firefighter should be left alone in a structure fire. He did, however, cite an example of when this may happen and gave alternatives for the student to consider. Assume you are the department chief. Give your solution for this problem.

The Mission of Ventilation

TOPICS

- The mission of Ventilation
- The reason for ventilating structures
- The basic types of ventilation
- Natural ventilation
- Natural phenomena associated with natural ventilation
- Mechanical ventilation
- Devices used with mechanical ventilation
- The type of ventilation most appropriate for specific circumstances
- When to ventilate
- The responsibilities of Ventilation

The mission of Ventilation is to remove or channel the products of combustion mechanically or naturally from the fire structure into a nonthreatening area. Ventilation, the last of the "Big Four" essential functional sectors used at structure fires, lessens the burden on working crews at fires by removing heat and smoke from working areas and improving visibility. If done correctly, ventilation also provides additional survival time for trapped or fleeing occupants.

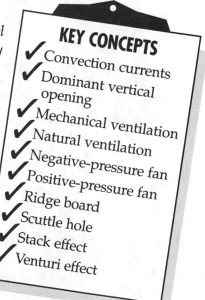

KEY CONCEPTS

- ✓ Convection currents
- ✓ Dominant vertical opening
- ✓ Mechanical ventilation
- ✓ Natural ventilation
- ✓ Negative-pressure fan
- ✓ Positive-pressure fan
- ✓ Ridge board
- ✓ Scuttle hole
- ✓ Stack effect
- ✓ Venturi effect

THE RELATIONSHIP BETWEEN COMMAND AND VENTILATION

Ventilation is an art and, in my opinion, an acquired skill. Good truck officers have certain instincts that make them successful in ventilation operations. Even seasoned fire officers who have little experience on a truck have

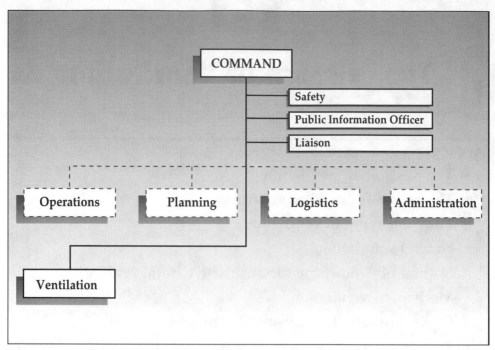

Figure 14-1. The relationship between Ventilation and Command.

difficulty in determining where, when, and how to vent a structure being attacked by fire. Truckies don't just "gut" a fire out (although opening a gable roof with a fire ax is the most consistently demanding work on the fireground). Some logic—not just brute force—must be behind the truckies' actions.

Just as an improperly applied hose stream can push fire to unaffected areas of a structure, a ventilation hole improperly located on a roof can "pull" heat and smoke to unaffected areas within the structure. Command must understand the principles that apply to and affect ventilation. More importantly, Command needs well-trained and competent officers.

With the incident management system, Command need not focus on ventilation but can do the following:

- logically place ventilation somewhere on his to-do list;
- assign ventilation to the most appropriate unit on the scene when it pops to the top of the list; and
- be confident that the sector officer assigned to ventilation will accomplish the task with expedience, enthusiasm, and expertise.

It all goes back to training and experience. Some smaller departments do not have the luxury of having permanently assigned truck officers. (Many departments do not even have trucks.) This does not mean they need not vent single-family residential structures involved in fire. What it means is that they need to cross-train their officers in numerous fire techniques.

When Command assigns ventilation, he has in the back of his mind a picture of what he believes needs to be done, whether it is to open the roof or only the windows. If Command has the time and knows how he wants the building ventilated, he should tell the truck officer when making the assignment.

Command to Truck 13: You're Ventilation. I want you on the roof!

This transmission leaves little doubt as to
- who is to vent,
- how ventilation is to be accomplished, and
- what tools should be brought up to the fire.

I do not believe Command need tell Ventilation where to put the hole or place the fan. This officer gets paid more than the firefighters on the truck because he knows his job. However, relief officers, newly promoted officers, or officers on trade may ride the seat of the truck. In these instances, the time Command spends giving a little direction may save much time and effort later—and even parts of the structure. When a less-experienced officer is riding a truck, I may take the time to tell him how and where I think the structure should be vented. That's part of my job as Command. If the structure is not vented at all, vented poorly, or vented in a manner that is counterproductive, that reflects not only on Command but on the department as a whole. Let's look at this a little closer.

SCENARIO

Figure 14-2. The second floor of this residential structure is heavily involved in fire. This fire has vented intself.

Command is at the scene of a working fire in a single-family house with heavy involvement on the second floor. Command assigns the lieutenant responding on the truck to be Ventilation; the lieutenant has seven years of experience.

Command to Truck 7: You're Ventilation.

That is the only transmission Command makes concerning ventilation. As Command is standing at the command post, he sees out of the corner of his eye the truck company walking toward the building with ground ladders and the chain saw. In his mind, he now knows the truck crew is going to vent the building, and by the means he had in mind (opening the roof). A hole at the ridge board of this structure, preferably over the attic stairs, is the best location for the hole. A hole approximately 36 inches × 36 inches (approximately two rafters square) would be adequate. Command has worked with this officer for years and knows the officer will vent as directly over the dominant vertical channel as possible, and with a hole two rafters square. Command need not walk up the ladder to ensure the location of the hole or its size.

Now let's look at another scenario.

SCENARIO

Figure 14-3. Light smoke is showing from the first floor of a structure.

Command is at the scene of a small kitchen fire on the first floor of a two-story house (above). He knows that a newly promoted officer is riding the truck today. Many things are going on in front of the house. The occupants are concerned about their pet dog. The first-in engine took a hydrant that was frozen and can provide only booster-tank water. On top of that, the second-due engine is held up by a train three blocks away. Command knows several things:

- At the present time, the fire involves only the kitchen near the stove and some cupboards directly above the stove.
- The fire is small and should be extinguished by less than the 500 gallons of water on the scene.

- Water must be put on the fire quickly, or it may spread to other parts of the house and beyond the capability of the 500 gallons of water available.
- The truck coming has a new officer with three members. Command makes the following assignments.

Command to Engine 5: You're Attack.

Command to Truck 5: You're Ventilation.

As soon as the transmission is completed, the occupant starts telling him about the puppy still in the house. Command has no time to give additional instruction to the new truck officer. In Command's mind, two things will happen:

- The attack officer will take a 1½- or 1¾-inch line through the front door and push the fire out the back, and
- The truck company will set the positive-pressure ventilation (PPV) fan and *also* push the smoke out the back.

As Command is explaining to the rescue squad officer where to look for the puppy, he sees out of the corner of his eye the truck company approaching with a ladder and the chain saw. This time, he tells Ventilation that ventilation is to be accomplished with a PPV fan and not by punching a hole in the roof.

Would Command be responsible if he did not see the truck company coming up with the ladder and if the crew had started to open the roof? No. To be responsible, you must have "knowledge" that a specific situation exists or an action is being taken.

There has to be trust between Command and all of his sector officers. The ventilation sector will affect the outcome of the incident, as will attack, search, backup, exposure, and so on. The problems may be different, but the trust must be the same. If Command has information that will (1) aid the officer, (2) take the mystery out of the assignment, and (3) not lead to micromanagement, he should give it.

WHY VENTILATE?

I myself have asked, Why ventilate? at many fires. The following *basic* physical facts make ventilation (for the most part) a necessary aspect at every structure fire. As "stuff" burns, the combustion process produces heat and toxic gases. The gases, generally referred to as smoke, become lighter than the surrounding air and rise until they meet an obstruction. The smoke then will travel horizontally until it can rise again. If contained by walls and a ceiling, smoke will bank down through an opening such as an archway, window, or door until it can rise again. This is a blessing for firefighters (and everyone else). Think of what the world would be like today if heated gases from fire (hostile or friendly) were heavier than air! We wouldn't be burning for too long—or even more correctly, everything would have burned long, long ago.

As the by-products of combustion rise, they can either be
- let out at the top of the structure through a hole provided,
- pushed out of openings in a more horizontal fashion, or
- left to collect at the upper levels of the structure to either accumulate further or dissipate naturally.

Therefore, we ventilate for the following reasons:
- *To stop the damage that smoke and the other products of combustion do to "stuff" inside buildings and the building itself.* Smoke and water are the leading causes of fire damage. Smoke coats "stuff" and leaves an unpleasant odor.
- *To assist escaping or trapped occupants, including pets.* Trapped and fleeing occupants stand a better chance of surviving if quick, properly placed, and channeled ventilation takes place. Smoke, which includes carbon monoxide, is the leading killer at fires.

At fires where occupants are still in danger and exposed to heat and smoke, ventilation should be one of the first actions taken. This is a difficult task, and it may cause the fire to spread to other parts of the building. However, it should always be considered when lives are in jeopardy.

- To make our job easier. Ventilation helps visibility. When searching, it is much easier to look for victims than to feel for them. The quicker ventilation can be completed, the quicker visibility returns to the area—making the operation safer and facilitating a more complete search. Ventilation allows us to enter areas that have an overabundance of heat. I have been on the top of a stairway unable to advance because of the heat. With every passing second, I would be forced back down the stairs. As soon as those truckies opened the roof, it was as if a blanket had been lifted

Figure 14-4. Firefighters working in a partially vented room. Proper ventilation can improve visibility, making the firefighters' task easier.

off me. Almost instantly, I was able to advance. Ventilation makes the work area more comfortable. In the hot summer months, I have placed electric fans in windows on fire floors to provide horizontal ventilation. During overhaul, this can mean the difference between "okay" and "spent" crews. Adequate ventilation during overhaul can reduce the time needed in rehab. Although I haven't seen studies to substantiate it, there must be a direct correlation between good ventilation during the latter phases of firefighting (overhaul, salvage, and button up) and lower vital signs during rehab.

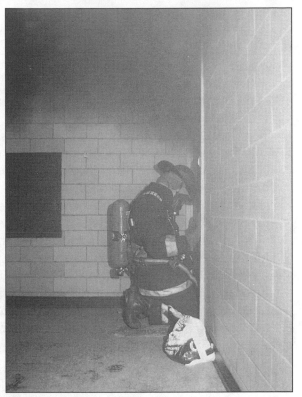

Figure 14-5. Quick ventilation can minimize damage such as that shown on the walls here.

TYPES OF VENTILATION

There are two basic types of ventilation: natural and mechanical. Both are effective under certain conditions and ineffective under others. The ventilation group officer determines which venting method should be used unless Command or Operations orders otherwise.

Natural Ventilation

In natural ventilation, the products of combustion are removed or channeled from the structure using natural processes, such as the following:

- *Convection currents.* This physical process allows smoke and other products of combustion (gases, burned and unburned fuels, for example) produced by a burning fire to rise. Smoke and other products of combustion will naturally be drawn out of the structure if an unobstructed path is provided.
- *Wind.* A significant factor in firefighting, wind will move smoke and other products of combustion throughout a structure. It will affect smoke movement in direct proportion to the following:

—the wind velocity

—the height of the building (normally, ground-level winds are less force-ful than winds at higher elevations, due to the friction of the ground), and

—the size and arrangement of openings. The larger the opening, the more movement. Equal openings on the windward and leeward sides create the most movement.

- *Stack effect.*[1] Present in high-rise buildings and more prevalent in very cold or very hot weather, this phenomenon—a flow of air in vertical shafts created by the difference between the air densities and pressures on the inside and outside of the building—helps to remove the products of combustion from the structure. If the inside and outside air temperatures are equal, there is little if any air movement. If the outside temperature is less than the inside temperature, the movement flows from bottom to top openings. However, if the outside temperature is greater than the inside temperature, the air will move from the top to the lower portions of the structure. In addition to inside and outside temperatures, building height, the air tightness of exterior walls, and air leakage between floors are factors in the stack effect.

Methods Used in Natural Ventilation

Open windows to allow the wind to push the smoke out of the structure. Normally this process is used at small fires where the production of smoke and heat are small. The process includes first opening the top portion of the windows on the lee side and then opening the lower portions of windows on the windward side across from the fire area (Figure 14-6). This method can be slow, depending on the wind speed and access to the fire area (basement or areas not in direct line with the wind). Additionally, it may take consider-

Figure 14-6. Air enters the windward side; smoke, heat, and gases exit the leeward side.

able time to remove smoke that is above the window line if the wind speed is slight to moderate. Since most of the fires to which we respond are handled by one line or less (the fire was extinguished with an extinguisher or before the fire department's arrival, for example), this process is used a large part of the time, particularly in rural America, where truck companies are not prevalent.

Make a hole in the upper level of the structure (normally the roof), allowing convection currents to draw out smoke and other by-products of combustion. Normally a truck company (although any company with an ax can chop a hole) will get to the roof of the structure and cut a hole to let out the heat and smoke.

Rules of Thumb for Venting a Roof

When venting a roof, cut the hole as directly over the dominant vertical channel as possible. Communication between Ventilation and Attack may be required. If the heat has an unobstructed path to the roof (if there is no ceiling or other horizontal obstruction such as in an attic floor, cockloft, or one-story warehouse fire), place the hole as directly over the fire as possible. If the fire is in an area where the path is obstructed, make the hole over whatever vertical avenue (called the "dominant vertical channel") the heat is taking. If the attack crew punched a hole in the ceiling with a pike pole, make the ven-

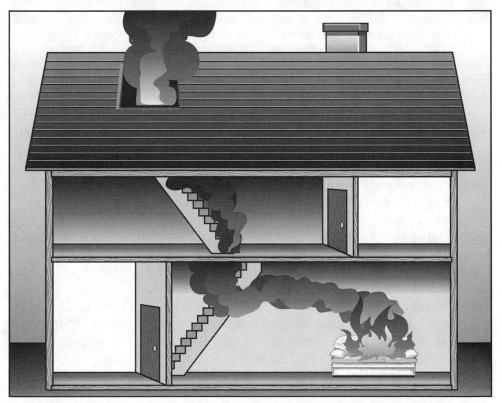

Figure 14-7. At this first-floor residential fire, smoke from the sofa fire rises until it finds the "dominant vertical channel."

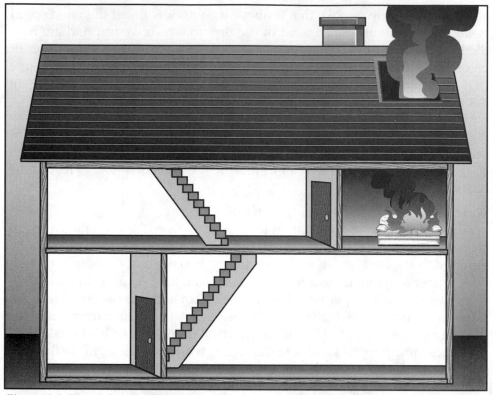

Figure 14-8. Fire on the second floor. The attack crew pulls the ceiling in the bedroom to let the smoke up into the attic, where a vent hole is directly over the fire.

tilation hole directly over that hole. If the heat is allowed to move up natural openings such as skylights, attic stairs, and scuttle holes, the hole on the roof should be directly over that natural hole (or dominant vertical channel). When this cannot be done, the heat and smoke's moving up one opening and horizontally across and upward to another opening may cause additional damage at the stairs up to the attic. The traditional tool for opening a hole on the roof has been the fire ax. Opening the roof with an ax takes strength and endurance. The newer tool used in topside ventilation is the carbide-tipped chain saw, which is similar to the normal home chain saw except that it has greater running speeds (which is helpful for cutting through nails and other roofing materials) and saw teeth with carbide-steel tips for strength.

Each tool has its place on the roof. I have found that the chain saw is quicker and safer for use on the flat commercial and newer plywood roofs. However, being of the old school, I am not yet convinced that a truck crew cannot cut a hole as quickly and more safely on an older gable roof with a fire ax. It takes considerably more time to carry the saw up the ladder and start it than to carry an ax up the ladder with you. Additionally, it is not safe to use a chain saw without a roofing ladder. The majority of the gable roofs in my jurisdiction can be opened without the use of a roofer. You can use an ax without a roofer. You can't use a chain saw without a roofer. Look at the time lost carrying, raising, and

Figure 14-9. A carbide-tipped chain saw. (Photo by author.)

positioning a roofer. The fire ax has its place on a roof.

Don't let yourself be cut off by the heat (and possible fire) that comes out of the vent hole. Keep yourself between the ladder and the hole. The hole for a normal house fire should be cut as near the ridge board as possible. It should be a minimum of two rafters square. If the rafters are on 16-inch centers, then the hole will be approximately 32 inches × 32 inches. If the rafters are on 24-inch centers, the hole will be about 48 inches × 48 inches.

To use the stack effect, open the door or hatch above the vertical opening in the high-rise building (normally a stairway or elevator shaft). Open the door on the fire floor to allow the smoke to be pulled by an effect similar to the venturi effect, and then make an opening in the lowest level on the shaft (normally the street level or lobby door). The stack effect will pull air from the lower level up through the shaft and out the top. In doing so, the flow moving past the open door on the fire floor will draw smoke from the fire floor out and into the shaft and up.

Mechanical Ventilation

Mechanical ventilation, the use of mechanical (along with natural) processes to remove smoke and other products of combustion from the structure, normally is accomplished by one of three means.

Fans

Probably the most common method of mechanical ventilation, fans are of two basic types—negative pressure and postive pressure.

Negative-pressure fans are the older electric fans carried on the trucks or other apparatus. (In some departments, they may have been put on the apparatus floor to make way for the new, more powerful positive-pressure ventilation fans.) Negative-pressure fans still have a place on apparatus. They normally are hung by hooks on window moldings, ledges, or door

Figure 14-10. A negative-pressure electric fan. (Photo by author.)

frames and pull smoke from inside the structure to the outside, thereby creating a negative pressure inside the structure (or area to be vented). They generally are smaller and can move only approximately 5,000 cfm (cubic feet per minute) of air, although larger "smoke ejectors" can move up to 15,000 cfm. The smoke ejectors could move quite a bit of smoke but are heavy and hard to maneuver. Additionally, you can't just plug them into any household electrical socket. Special generators or adapters are usually required for their use.

Positive-pressure fans are more recent innovations in the fire service. I'm

Figure 14-11. A positive-pressure fan. (Photo by author.)

not sure the final chapter has been written on their use. They are very effective for ventilating a structure if used correctly. They average about a 20,000-cfm capacity. The fan is placed outside the structure and forces air inside, creating a positive pressure of air within the structure. If there is no avenue for this forced air to leave the structure, the air will back up and seep out of small natural openings, as well as back out of the same opening into which it is being forced. This can be very dangerous. However, if an opening to the outside is made in an area where the products of combustion are, this positive air pressure will force the smoke out of that opening.

This method has several advantages:

- Smoke from almost anywhere can be channeled inside a structure to almost any exhaust location. If done correctly, a fan set in front of the house can move smoke from a first-floor bedroom with no windows, down a hall, and out another room without filling other areas with smoke. Additionally, properly placed and channeled PPV can temporarily move smoke and heat away from trapped or fleeing victims to less threatening areas.
- Smoke from a specific fire area can be forced out a window in that area without being dragged throughout the house. You can virtually "localize" the ventilation process.
- It helps to locate small hidden fires. If looking for such fires, start the fan and stand back. Fanning the fires with a blast of air will intensify them. We have used this technique many times when overhauling fires.

Some distinct precautions, including the following, must be taken with this form of ventilation. It is not a ventilation panacea. If not used properly or at the correct time, it can worsen fire conditions and possibly harm fleeing victims.

- Do not start the fan until you know where the seat of the fire is. When these fans were first introduced, one of the selling points was that you could turn this fan on while advancing toward the fire. I question this concept. If you pull up and have considerable smoke and no open flame coming from the front of the structure and blindly turn on the PPV fan without knowing the location of the fire, you will fan this fire with additoinal fresh air and push it toward any openings in the building. It may be nice to have fresh air at our backs while advancing toward a fire, but at what cost? Find the fire, then turn on the fan. The only exception to this is if the location of the fire is clear and you know where the heat and smoke will be exiting. In those instances, you can certainly turn on the fan even prior to advancing. When using the PPV fan, consider the same rule of thumb as for a fire attack: If the fire is in the rear, put the fan in the front; if the fire is in the front, put the fan in the rear.
- If you prematurely turn on a PPV fan prior to doing a primary search, you may be pushing heat and smoke on a downed or trapped victim.

SCENARIO

You respond to a single-family house fire at 1730 hours. As you pull up, you observe heavy smoke coming from the front door. You order an attack sector with the remainder of your crew and assign ventilation to the truck company pulling up. While your crew masks up and pulls a line, the truck crew puts the PPV fan in the area of the front door and turns it on. The fire appears to be in the rear of the living room and spreading to the kitchen area. The truck crew officer goes to the back door and opens it to provide an avenue for the smoke. What if the occupant had discovered the fire and was attempting to flee the house through the kitchen when she was overcome by smoke? You may have just fried her!

If in this scenario the line were taken in through the back door (if the fire's in the front, the first line should go in through the back—attack from the unburned portion toward the burned portion, and attempt to push the fire back to its point of origin) as it should have been, would you have the attack line advancing toward a running PPV fan? Which would win—the fan or the fire stream? Wait until you find the fire and knock it down, then turn on the fan.

HVAC

Heating, ventilation, and air-conditioning (HVAC) systems present in newer commercial buildings such as high- and low-rise office buildings may be used for ventilation. These systems are quite sophisticated and may require than an engineer be on duty 24 hours a day to operate them. HVAC systems can pull any and all products of combustion from the fire area to the outside, or they can pull smoke and heat throughout the entire structure. Preplanning is the key. Know your buildings.

Fog

A fog stream is often overlooked as a means of emergency ventilation. A 1½-inch line flowing 60 gpm at 100 psi nozzle pressure can produce from 10,000 to 30,000 cfm of air movement,[2] depending on the size of the opening and the application method. This is one of those rare times I advocate using a fog stream inside a structure. Since the stream is directed outside an open window, not much steam, if any, is generated. Use this form of ventilation only in rare instances, such as to facilitate a search. I used it when I was a lieutenant assigned to #1 squad. We were at a single-family, 1½-story residential structure fire conducting a search on the second floor. The first thing one of my crew members yelled to me when we made the second floor was that he had found an empty crib. About that time, another member yelled that he had toys on the floor. We now knew children might be inside. A company came up behind us with a hoseline. I had the team quickly vent the area with a fog stream. Almost immediately, smoke conditions improved. Now, instead of feeling for victims, we could look for them. Luckily, we found no one. This form of ventilation can quickly clear a room if you need to look for someone.

Figure 14-12. Firefighter venting using a fog stream.

The Procedure

Open or break out an entire window. Put the nozzle on 60° fog, and stand back from the window. Open the nozzle, and move forward or back until the majority of the stream is flowing out the window. Some problems are associated with this means of ventilation.

You can pull heat and fire into your area of operation if you are not careful. It may be wise to close the door behind you (if the line does not stop you) to keep any heat, fire, and smoke out of your area. Backup should be notified to take a more defensive position between you and the original fire area.

Attack to Backup: We're going to have to vent a room on Division 2, Sector C, with fog. Keep the fire moving down the hall on us.

You will cause additional water damage to the structure. It's hard to throw a perfectly round stream through a rectangular window. To cover as much of the opening as possible, some water will be thrown on the floor and walls. If you must use this method, use it only until you can see; then have the stream shut down.

WHICH TYPE OF VENTILATION TO USE?

Which type of ventilation should be used depends on the following factors:

• *The location of the fire.* If the fire is on the ground or second floor of a residence and it has not spread to the attic, mechanical ventilation, using fans, is probably called for. The proper use of PPV is proving to be the most effective means of selectively channeling smoke to or from areas. If the fire has entered the attic or cockloft or the structure itself (inside floor or wall assemblies), you may need to open the roof. If this is the case, then, in my opinion, you don't want to turn on the PPV fan.

- *The sectors assigned.* There will be times when fans will be counterproductive to other working sectors. When this is the case, topside ventilation may be considered. In some cases, simply opening windows may help for the short term.

- *The stage of the fire.* If the fire is at the flashover stage, I don't advocate putting the PPV fan in front of a door. If the fire is in the high-temperature, low-oxygen smoldering stage, you don't want to open the door and turn on the fan (that is, if you could turn it on prior to the "boom").

Experience is the answer. By watching what works and what doesn't, you gain an understanding of the above-mentioned principles and concepts discussed in this text and others.

There are a few "keys." When I was a young lieutenant, I tried to fit all the "signs and symptoms" of firefighting into neat little packages and came up with these little cure-alls: "If this, then that." Sadly, it isn't that simple. Fire situations are dynamic, complicated, and unique.

WHEN TO VENT

Ventilation should be started as soon as possible. Some factors must be considered when determining where ventilation should appear on Command's to-do list.

Ventilation begun too soon or in the wrong location can cause the fire to spread to unwanted areas and intensify. I was at a fire recently where a firefighter freelanced and broke several first-floor windows from the outside on Side 4 of the structure, giving additional oxygen to the fire. Within seconds, the fire flowed toward that opening. If backup had not been doing its job, we might have cut off several firefighters doing a search on Division 1. That was the wrong time and way to vent that fire. The proper time to ventilate a building is dependent on the following.

- *The stage of the fire.* If the fire is in the incipient, free-burning, or high-smoke-producing stage, ventilation should be relatively high on Command's to-do list. Quick ventilation, probably the second or third assignment made (as long as search is not a major concern), will be required. If the fire is in the flashover stage, more heat is being produced. Normally there is plenty of oxygen in the area during the flashover stage. Topside ventilation probably will be necessary. In the high-temperature smoldering fire (backdraft potential), ventilation is the first item on the to-do list.

- *The extent of the fire.* Is there a small fire in a single room, or is the entire floor (or are floors) in flashover? If only a part of a floor (excluding the attic) is involved, ventilation should be high on Command's list. The rule is, the more involvement (or flashover), the higher the place ventilation takes on the to-do list, up to the point of total involvement. {If the fire is in the high-heat, low-oxygen smoldering stage (backdraft potential), ventilation must be the first item on the to-do list.}

- *The location of the fire.* If the fire is in an area where the smoke and heat being produced are or could be affecting the efforts of those attempting to flee, ventilation must be very high on the to-do list. Conversely, if the by-products of the fire are not affecting the occupants, the structure, its contents, or firefighting efforts (such as in an attic fire), ventilation can be placed lower on the to-do list. A rule of thumb that could be applied is that in an occupied dwelling, the lower the level of the fire involved, the higher the priority ventilation takes. (There is more area above that can be affected.) Some factors such as location of victims, time of fire, and extent of involvement may change this rule, but it works.
- *The occupancy type.* For the most part (unless otherwise indicated), this text deals specifically with single-family structure fires. If a fire involves a plant that manufactures explosives, ventilation may not be a concern (depending on a few key items). Ventilation in a public-assembly occupancy will take precedence over many other functional sectors. Ventilation in a vacant single-family structure that has had previous fires would have a lower place on my to-do list than if the fire were on the first floor of an occupied home at 1530 hours on a Saturday afternoon in the wintertime. Know your buildings!
- *The other needs of the incident.* It's hard to make blanket statements about the priority of specific sector assignments at incidents. There are so many variables. Keep in mind the four priorities of Command discussed in Chapter 2: firefighter safety, civilian safety, stop the problem, and conserve property.

What may be key at a fire in a building one day may be insignificant in a fire in the same building three weeks later. I hope several key phrases and concepts that I have repeated (at times ad nauseaum) have stuck and will stick with you for years. One of those is "look at the picture." Speculation while responding is fine and many times beneficial. (Good, up-to-date pre-plans are worth their weight in gold. However, it all changes when you pull up in front of the structure. If there is no evidence of a fire and no one is waving you down with breathless anticipation, then the location of the gas meter and the name of the keyholder are worthless pieces of information.) If you pull up to a four-family apartment building and have only very light smoke showing from the rear door of a unit, ventilation may be relatively low on the list. However, if you pull up to the Valentine Theater and have people hanging from the second-floor windows and occupants being trampled to death trying to escape the smoke, then ventilation should pop to the top of the list.

Look at the picture! If a specific rescue or a quick knockdown can be made, then it may take precedence over the need to ventilate immediately. Keep in mind that the biggest hole in the roof would help only a little if the fire were allowed to burn uncontrollably. You have to get in and put the fire out to be effective.

- *The personnel and equipment on-scene and responding.* You can't expect your crews to do it all. If your department is like mine, personnel normally is not a concern. I have a minimum of 103 personnel on duty 24 hours a day, 365 days a year. However, I realize that my type of department is the exception and not the rule. Most departments need to rely on mutual aid to fight even the normal working fire in a single-family residence. Additionally, the first few minutes on the scene do not provide the 15 to 25 firefighters I get (normally within the first three minutes). Today I was reading an article in *Fire Engineering* concerning a firefighter who became trapped on the third floor of an exposure.[3] The author was discussing a fire in a large three-story multifamily apartment. He said the incident commander "recognized he had a working fire involving the top floors and cockloft area." The IC special-called an additional "third" truck company and ordered the truck company to split into two teams of three to check the exposures. Split the truck company into two teams of three! I don't think so. I've never seen a five-person truck crew in Toledo,[4] let alone a six-person crew. How do we survive with three-person truck companies? We do what we can. My point is that Command must look at the picture, develop a to-do list, and then assign on-scene crews to accomplish the tasks. Know your crew's capability. If you know you have about three minutes to vent the roof if you expect to find viable victims inside, and the normal time to open a roof for a truck crew is six minutes, go to plan B.

THE RESPONSIBILITIES OF VENTILATION

As noted several times (I believe repetition is vital to learning), **sector officers are responsible for the safety of their crews.** When venting a roof, this is a key concept. The officer assigned to ventilation needs to detemine the stability of the roof (which is an acquired skill). Sagging rafters, sponginess, and warm or hot spots are indications that the fire has entered the attic or cockloft area. In the winter, specific and uncharacteristic melting of snow or ice is a clear indication that the fire has entered the space below you. Once on the roof, crews must not cut themselves off from their means of egress. Place holes with the wind, and you're way off the roof on your back.

A newer concern with ventilation is the use of SCBAs while venting. Mandatory mask policies dictate that SCBAs be worn while venting. I am glad this policy was adopted after I made chief. It's hard enough to cut a hole, let alone trying to do it with an SCBA on and in service. However, if a department has a mandatory mask policy, the sector officer is responsible for seeing that the members under his command follow all safety procedures. We have had a rash (at least four occurrences in the past two or three years) of

PPV fan blades fracturing and being thrown out of the fan housing. Luckily, no members (or civilians) were hit with the flying blades. If you're having a problem with ventilation tools, including cutting tools, it is up to the officer to make the administration aware of the problem and to provide the safest working area possible.

The ventilation officer is responsible for ventilating the assigned structure or area. If Command can't take the time to give detailed instruction on how he would like ventilation to be handled, it is up to Ventilation to choose what he believes to be the most appropriate means of venting and then to direct his crew while they accomplish that task. It may be as simple as rolling the

Figure 14-13. Ventilation crews operating on a roof.

Figure 14-14. A truckie using a carbide-tipped chain saw. Notice the use of a roofing ladder for adequate footing.

PPV fan up to the front door and turning it on when Command gives the word. Or, it may be as complicated as channeling smoke through a manufacturing plant away from fleeing workers.

Ventilation must keep Command informed. If Ventilation is on the roof looking at a 48-inch by 48-inch hole and the attack crew is still screaming that it can't make the second-floor stairs, Command needs to know that. Once the hole has been opened and smoke is coming out, give the appropriate benchmark. When Command gets reports contrary to what he has been told, he can start to pinpoint the problem.

Endnotes

1. O'Hagan, John T. *High-Rise/Fire and Life Safety.* 1977. Fire Engineering Books: Saddle Brook, N.J., 155.
2. Clark, Wm. E. *Fire Fighting/Principles and Practices.* 1974 (2nd printing 1976). Fire Engineering Books.
3. Pressler, Bob, "Firefighter Trapped," *Fire Engineering,* Dec. 1995, 56.
4. A three-person crew on a truck is a minimum in Toledo. We rarely exceed that number.

REVIEW QUESTIONS

1. What is the mission of Ventilation?

2. What may occur as the products of combustion rise?

3. What are the reasons for ventilating?

4. What are the two basic types of ventilation?

5. What is natural ventilation?

6. What three natural processes assist in removing or channeling the products of combustion from a structure?

7. What are the methods for natually ventilating a building?

8. Define mechanical ventilation.

9. What are the most common methods of mechanical ventilation?

10. The type of ventilation used depends on what three factors?

DISCUSSION QUESTIONS

1. Define dominant vertical channel, and explain its relation to topside ventilation.

2. Justify keeping a negative-pressure fan on a ladder truck.

3. Compare and contrast the use of an ax and a chain saw for topside ventilation. When do you think each should be used? Why?

The Mission of Exposure

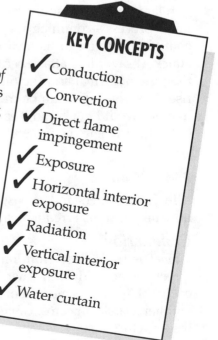

KEY CONCEPTS

- ✓ Conduction
- ✓ Convection
- ✓ Direct flame impingement
- ✓ Exposure
- ✓ Horizontal interior exposure
- ✓ Radiation
- ✓ Vertical interior exposure
- ✓ Water curtain

*T*he mission of Exposure is to prevent the spread of and to extinguish any fire in the assigned area. This means gaining access to the exposed building or area to check for and mitigate any fire spread into the assigned area or exposure.

THE EXTINGUISHMENT PROCESS

The extinguishment process has three components:
- to confine the fire,
- to control the fire, and
- to extinguish the fire.

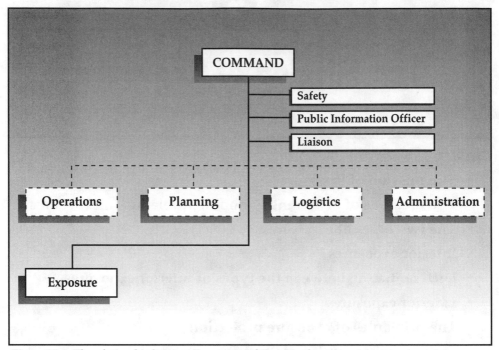

Figure 15-1. The relationship between Exposure and Command.

THE EXTINGUISHMENT PROCESS AND EXPOSURE PROTECTION

In large part, exposure protection involves the first part of the extinguishment process—confining the fire. Any fire attack strategy should include confining the fire to its area of involvement at the moment of the initial attack. (I say initial attack because there may be a significant lapse between the time of arrival and the time the initial attack has commenced. At high-rise fires, this lapse can be more than 20 minutes. An astute incident commander would not commit forces to the area where the fire presently is but to the area where the fire will be when the water starts to flow.)

Heat Transfer

To fully understand exposure protection, you must have knowledge of how heat is transferred (spreads). There are four methods of heat transfer:

Conduction
Heat is transferred from one molecule to another as molecules come in direct contact with each other. Heat always flows from hotter objects toward cooler objects. This method of spread is indicative of spread from one area to another via steel beams, cable, wire, and even nails in hidden areas within the structure.

Direct Flame Impingement

The flame of a burning object comes in contact with another object—most significant in incipient fires. Impingement also affects fires in the free-burning and flashover stages. Direct flame impingement is a factor in fire spread to exposed buildings.

Radiation

The transfer of heat by radiation is in the form of electromagnetic waves that travel in straight lines from the source of the fire. When these waves strike the surface of a material, it causes the atoms within the molecules of the material to move faster, creating heat. The closer to the source, the stronger the electromagnetic waves, and hence the more radiated heat. Radiation is a concern with exterior exposures and to a lesser extent with interior exposures.

Convection

Convected heat moves through a circulating medium. Watch a campfire. The plume of smoke and heat rising up from the fire is caused by (and is) convection. Convection is responsible for most of the fire spread inside a structure.

Although radiant heat is released from a convection current, the major spread is upward. Convection current (and smoke that rides these currents) will travel upward until it dissipates or meets an obstruction. If an obstruction is met, the current will travel horizontally[1] until it finds another way up. If no outlet is found, the heat will bank down until it finds a way up or dissipates.

TYPES OF EXPOSURES

There are two types of exposures—interior and exterior. Protecting these exposures is discussed later in this chapter.

Interior Exposures

We don't usually refer to interior fire spread, from room to room or floor to floor, as interior exposures.[2] We generally enter the structure and make an "unconscious" effort to stop the spread of fire and keep it to the smallest area of involvement. Interior exposures are areas within a structure that could become threatened by the spread of fire. Interior exposures are most frequently thought of as fire spread from room to room, including hallways, although fire spread can also be from floor to floor.

There are two basic forms of interior fire spread, and thus two forms of interior exposures:

Horizontal Interior Exposures

Fire spreads horizontally, basically from room to room (or from furniture to furniture within a room). This form of spread is caused mostly by radiation and direct flame impingement.

Vertical Interior Exposures

Fire spreads to upper portions of a room or from floor to floor, normally by convection currents.

Figure 15-2. A fire with exterior exposure problems. Where would exposure be on Command's to-do list?

Exterior Exposures

These exposures are items of value outside the structure or area of origin that are or could be threatened by fire or the products of combustion. Exterior exposures are normally, but not exclusively, structures. Command or department procedure determines what constitutes items of value. Other homes (vacant or occupied),[3] vehicles parked in driveways, barns, sheds, and garages are considered exposures. {Questionable exposures would be plants and shrubs, toys (bikes and so on), and fencing. All of these could certainly have value, but their relevancy as it relates to dollar loss and the use of personnel is questionable. The insurance company that underwrites the individual property owner will determine the value of these sundry items.}

EXPOSURE PROTECTION

In general, the need for protecting exposures is dependent on the number of Btus being produced and the ignition temperature of exposed material. If the amount of heat being produced from the source fire is small, the need for immediate exposure protection is not great. If the heat from the source fire is significant, but the exposed material—such as a concrete wall—has a relatively high ignition temperature, the need for immediate exposure protection is not urgent. Interior exposures are affected mostly by convected heat (although radiant heat will ignite materials in an area more horizontal to the fire). This indicates that for interior exposures the fire tends to spread upward rather than horizontally. With exterior exposures, the converse is true. Radiant heat, the culprit here, travels in a straight line from the source. The radiant heat from the burning fire travels horizontally from windows, doors, and exposed wall surfaces to adjacent exposed surfaces.

Interior Horizontal Exposure Protection

To protect horizontal interior exposures, locate the seat of the fire and place a hose stream between the fire and its expected avenue of spread (sources of spread). Hallways work well as these cutoff vantage points if the room is well-involved. (By the way, simply closing the door to an involved room will stop horizontal interior exposure spread. Most home doors are rated for 20 minutes. Older homes do not have rated doors, but older doors seem to have more "mass," which is directly related to resistance.) For smaller fires, advancement into the room or fire area will help cut off the fire. Finally, the amount of convected and radiant heat being produced must be reduced.

Interior Vertical Exposure Protection

To protect vertical interior exposures, locate the source of the heat and stop or check the spread of heat by convection currents. As noted in Chapter 11, I prefer the indirect attack for interior firefighting. In this method of attack, a straight stream is bounced off the ceiling onto the burning object. An indirect attack begins to check vertical interior exposures. It is mandatory to get above the fire as soon as it is knocked down to check for vertical extension. (Checking for extension is covered in Chapter 16.)

Exterior Exposure Protection

To protect exterior exposures, it is necessary to understand that the majority of heat transfer to exterior exposures will be from radiant heat or direct flame impingement. To protect materials from both of these forms of heat transfer, you need to cool the exposed material. Cooling the surface of the exposed material with water is the most effective way to protect exposures from these forms of heat transfer. Apply a straight stream onto the upper surface of the exposure and let the water wash down over the face of the exposed surface. This flow of water helps to keep the material below its ignition temperature. Water normally does not flow upward. Direct your stream to the upper portion of the exposure.

I see crews (and officers) assigned as Exposure commit two big errors. The first is directing their stream predominantly onto the original fire, neglecting the exposure. If a vantage point for applying the stream to an exposure permits effectively hitting *both* the exposure and the original fire source, then it is permissible to *occasionally* hit the source. Remember that until the source is knocked down, radiant heat will continue to be generated. A balance must be maintained between protecting the source and the exposure. Exposure must remind crews to keep cooling the exposure. If the source is burning to the extent that crews can't keep up with cooling the exposure, an additional line may be needed to continuously hit the source while the exposure line constantly cools the exposure.

The second most common error exposure crews make is not starting with a large enough line. Serious exposure fires call for a line larger than 1¾-inches. A 2½-inch preconnected line should be the first line pulled at a fire involving a significant exposure problem. What it lacks in maneuverability (which is not needed if initial line placement is correct) it more than makes up for in fire power.

The Roof Area

The exposure line may have to be moved back so that an effective stream can cool the roof in the same manner as the exterior side walls are cooled.

Finally, streams should be directed toward the other exposure items listed earlier in this chapter. Vehicles may be moved if the keys are available. Fences and other pertinent objects must be cooled. Placing water on exposed objects can keep them below their ignition temperature.

PRIORITIZING EXPOSURES

Interior Exposures

Normally, there is no doubt as to which interior exposure deserves your attention first. To a large extent, nature takes the guesswork away from you. For interior exposures, convection currents will demand that you check extension to the upper exposures first. If the attack sector takes the correct avenue into the structure—from the unburned portion toward the burning area—then he to a large extent is prioritizing the interior exposures. At fires in large occupancies such as nursing homes and public assembly buildings, the priority of interior exposures will need to be carefully considered. For now, I am concentrating on the normal house fires we experience day in and day out.

Exterior Exposures

Determining the order in which exterior exposures should be protected is a little more complicated. Several factors must be considered.

The Availability of Additional Companies

In urban departments, the availability of personnel usually is not a concern. These larger departments normally get two or three engines on every structure fire. In some areas in the outlying districts, the second engine may be five minutes away from the first-in engine; however, the first engine knows another is on the way. In rural America, this is not always the case.

Personnel on a structure fire in rural fire departments can range from plenty of personnel to one or two members on a single apparatus (while other members respond in their private vehicles to the scene). In these situations, the first-in officer must look at the picture as he considers the items below (among others):

Wind Direction

Wind probably is as big an influence as any on exposure protection. It will direct the course of interior fire spread. This spread, which follows the direction of the wind, will move and possibly intensify the fire. Wind also pushes direct flame and convection currents, which will affect spread. All things being equal, choose the exposure on the lee[4] side of the fire.

Wind Speed

There is a direct correlation between the speed of the wind and the effect on exposures. The more force behind the wind, the more the fire (and its spread) is affected.

The Proximity of the Exposed Buildings

All things being equal—life safety potential, economic value, wind direction and speed, and so on—the closest building to the source deserves the first attention.

Life Safety

Make every attempt to protect the greatest number of people first. If, for example, a source fire is affecting two exposures equally and one is a single-family frame home and the other a nursing home, direct initial efforts at protecting the nursing home.

Economic Value

Again, all things being equal (including the fire threat to the exposed structures), you probably will protect what you believe to be the structure of the most value. In most instances, this is a judgment call.

Occupancy Type

Is either structure vacant? Is the structure on Side 2 an unoccupied warehouse whereas Exposure 4 is an unoccupied museum? Again, a judgment call.

Pick your exposure, and then get a line between the source and the structure you're protecting. Cool the face of the exposed side as described above.

PROTECTING EXPOSED STRUCTURES

The usual picture that comes to mind when the word "exposures" is mentioned is a single-family residential structure. Perhaps in larger urban settings in the East, it may be another row house (attached by a party wall to the source structure). In rural America, it may be outbuildings such as barns and storage buildings. No matter what the picture, getting a line between the source fire and the exposed building is the first step in exposure protection.

Apply a stream of water with sufficient gpms (gallons per minute) to allow water to wash down the face of the structure to cool the exposed material below its ignition temperature.

To *totally* ensure that fire will not attack the exposure, you need to get inside the structure and make sure of two things:

Fire has not entered the structure. When I came on the job, every apparatus had a "water curtain"—a half-circle device with a 1½-inch coupling for a hoseline. Water would come out of an open pipe (just past the coupling) and strike the half circle, making the prettiest spray stream. The intent was to place this stream between the source and the exposure and let the stream do the rest. Well, it just didn't work!

First, it tended to straighten out the line in between the houses. We couldn't put a firefighter on the water curtain because fire burning to the extent that it would jeopardize nearby houses would endanger the firefighter as well. Second, and more to the point, it has been determined that a water spray absorbs only about 15 percent of the radiant heat produced by a fire. Another way of saying this is that water spray will stop only 15 percent of the radiant heat passing through it.

Don't let this fact confuse you with regard to what was previously discussed. Water allowed to wash down the face of an exposure keeps the surface of the material exposed below its ignition temperature. However, a spray of water thrown into the air between the source fire and the exposed surface will stop only 15 percent of the heat radiated. Glass, too, will stop only about 15 percent of the radiant heat being produced by a fire (the source). Radiant heat will pass through streams and glass and enter windows and other openings. This heat can and will warm up "stuff" like curtains and bedding above the ignition temperatures. Openings in the underside of eave soffits and other areas also can permit heat to pass into a structure.

Smoke and water are not entering the exposed building. Open windows in the summer and broken glass from

Figure 15-3. Good exterior exposure protection. Notice that two lines are on Exposure 2 and not one of them is putting water on the source fire.

heat and hose streams can permit smoke and water to enter the exposed structure. Even though fire may be kept from entering the exposure, smoke and water damage may occur. Steps must be taken to ensure that these two elements are not permitted into the exposure.

Normally after a line has been put into place to protect the exposure, the officer assigned to exposure should instruct the nozzleman and then send at least one member of his crew into the structure to check for fire, smoke, and water damage. It may be necessary to force entry. (If the possibility exists that heat, smoke, or water can enter or has entered the structure and the occupant of the exposure is not at home, I would advise forcing entry.) Check and recheck areas on the side of the fire source and attic. Close any open windows on the side of the fire source. If the window cannot be closed (if it has broken from the heat), you may have to place a hoseline inside the window to cool the area.

It is extremely important to get into the attic or cockloft. Heat rises. If the structure is of balloon construction, there may be spread up to the attic even through the fire came in through the lower levels of the structure. Once you have access to the attic, you must recheck this area often.

One question often asked is, Do we need to take a line into the exposure? If it is possible that fire has entered or could enter the exposure, then get a line. It's better to get one in now than to play catch-up later. The size of the line should be commensurate with the amount of heat being produced from the source, the proximity of the exposure, and the efforts outside on the side of the source fire. Big (or potentially big) fire, big water!

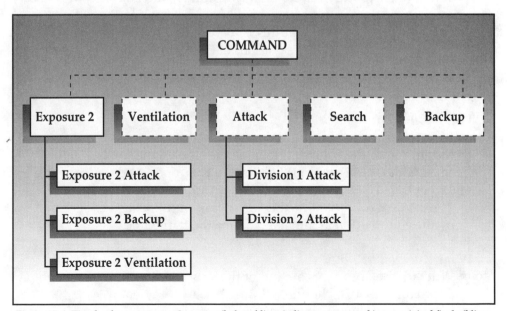

Figure 15-4. Fire that has exposures. Sectors wtih dotted lines indicate sectors working at original fire building.

EXPANDING THE SECTOR

As with most sector assignments, the sector officer may need to ask for more help. In one respect, the exposure sector has a unique problem. The sector officer asisgned to a specific exposure could become Command (for lack of a better term) of his own structure.

SCENARIO

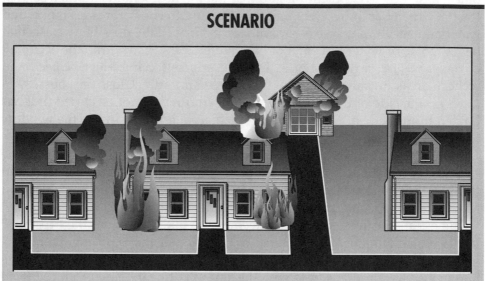

Figure 15-5. The original fire building (source fire) with exposures on all sides.

Let's say you are responding to a working fire in the center of the city. As you pull up to your staging area (Level I staging), Command assigns you to Exposure 2 (the exposure on the right side of the source fire). You can see from your staging area (and have learned from listening to the radio transmissions while responding) that this is a working fire. A large amount of flame is impinging on the upper portions of Exposure 2. You take a hydrant and lay in up to Exposure 2. You observe smoke coming from the eave area and roof. You can hear glass breaking.

What You Know
- It's a two-story, wood-frame occupied home.
- Judging from the color and force of the smoke, the fire has gained a small (although significant) hold on the the upper levels of Exposure 2.
- Direct flame is impinging on the upper portions of Exposure 2.
- A great deal of radiant heat is impinging on Exposure 2, determined by the sound of breaking glass (above the flame line).

What You Don't Know
- Whether the occupants are out of the structure (it is a winter night at approximately 0245 hours).
- The extent of the involvement inside.
- When the flame and radiant heat impingement will subside.

You have your crew pull a 2½-inch line and place it so that you can cool the roof and eave area of Exposure 2 and still occasionally hit the source fire. After the line is in place, you and another crew member force entry into Exposure 2. You find no smoke or fire on the first floor.

You yell to occupants. As you move toward the stairs, you smell smoke and then see light smoke on the second floor. (No detectors are going off.) As you get to the top of the stairs, you can hear crackling above you. You open two bedroom doors and find four family members, who are taken down and outside.

At this point, you know the following:

- There's a fire in the upper levels of Exposure 2.
- All occupants are safely outside.
- You are the officer of a four-person crew. One-half of your crew is keeping (or attempting to keep) fire that has not entered Exposure 2 outside, and the other half is you and one firefighter. At this point, you need to call Command:

Exposure 2 to Front Street Command: Be advised that we have fire in the upper levels of Exposure 2. All occupants are out. Exposure 2 will need two engines and a truck to handle this exposure.

Command must still be the only member on the fireground who asks for and receives additional equipment. It would not be prudent or within the scope of responsibility for Exposure 2 to ask for additional units from dispatch. (Exposure 2 may be on another fire frequency and may not have heard Command ask for additional equipment to be staged. Additionally, dispatchers must never send units at the request of anyone on the scene except Command. Command's task is to focus on the whole. The exposure is only a part of the whole. Someone needs to account for the entire scene.)

As Figure 15-4 indicates, Command has not exceeded his span of control at this fire. He still is responsible for only five subordinates. Under most circumstances, I would feel comfortable retaining Command at a fire of this magnitude without assigning an operations officer. However, I would let Exposure 2 run his fire. If he needs something, he can report to me physically (he probably will be out in front of these two houses and within 30 feet of me) or by radio. I would want to be kept informed, but I would want him to handle his exposure in its entirety. I probably would insist that he move himself and his sector officers to another radio channel if one is available. This will go a long way toward minimizing fireground confusion (units calling for attic ladders, more water pressure, or pike poles).

This type of incident presents an excellent opportunity for observing a member who is on the promotion list for chief officer or captain. You can allow the candidate to get his feet wet at what basically would be his own incident while you are only a few feet away.

Using the previous scenario as an example, it should be clear that, if the need arises, an officer assigned as Exposure (whether one exposure or all exposures) can have the same basic responsibilities as Command at a fire. If "his incident" (the exposure) enlarges, so must the exposure officer's responsibilities and, subsequently, his to-do list.

As can be seen in Figure 15-5, the original fire (source fire) is in a two-story frame home. There is heavy involvement throughout the structure. Exposure 2 already has begun to display involvement on Side 4. The garage in the rear (Exposure 3) also has involvement at the eaves line. Exposure 4 is the least problematic area at the current time. The winds are mild. If they increase, Exposure 4 may become a real concern.

As indicated in Figure 15-4, Command still has sole control of the scene. He has not yet assigned an operations officer. The dotted box indicates units assigned to the original fire building. It would be assumed from the chart that he has assigned an attack sector only to the original fire (probably exterior lines). Exposure 2 has blossomed into a first-alarm assignment. The original exterior line (Exposure 2 ext. line) operating outside, between the source fire and Exposure 2, is still in place. Exposure 2 now has an attack sector on Division 1 and Division 2. Both are protected by a backup sector and a ventilation sector (probably on the roof).

Exposure 3 (in the rear) is a two-story garage with storage space over the garage. An exterior line has been placed between the source fire and Exposure 3. A sector has been assigned to attack the fire on the second floor and in the attic, and Ventilation has begun to open a hole in the roof.

Exposure 4 is another home. The wind is helping to abate the problem in Exposure 4, but an exterior line has been stretched between the source and Exposure 4. A sector company (Exposure 4 Extension) is inside checking to see if the fire has entered Exposure 4.

Through all of this, Command has been in control and has not been responsible for more than five subordinates. Each exposure has an officer whose prime responsibility is to focus on the whole (in this case the particular exposure building) while Command sits back and makes sure every base is covered. Just for a second, think of how your department would have handled this type of fire without some form of the incident management system. A chief officer is literally running around the source building, *attempting* to make sure that all the exposures and their specific nuances are being covered. I can remember chiefs running back and forth around garages and jumping fences to cover all the exposure problems *effectively*. Doesn't it make sense to assign a specific officer to each exposure and allow Command to stand back and wait for the officers to present the problems to him to solve?

Exposure 2 to Command: We'll need the electric company to cut the drop to Exposure 2.

Command: Okay. I'll notify Dispatch.

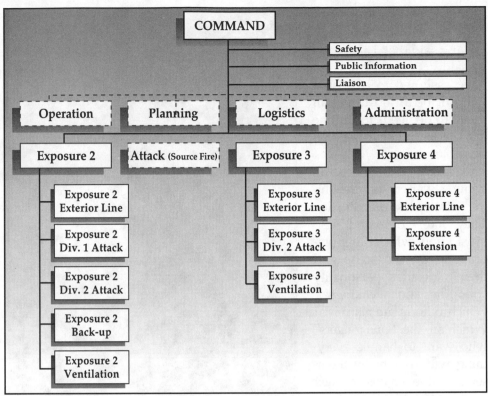

Figure 15-6. This flow chart represents the same scenario as in Figure 15-4 but shows the expansion. The original (source) fire now affects three separate structures.

Exposure 4 to Command: It got into the attic. I'll need an engine company for Exposure 4 Division 2 Attack and a company for topside ventilation.

Command: Okay. You'll get Engine 6 as Exposure 4 Division 2 Attack and Truck 13 as Exposure 4 Ventilation.

If Command had time to focus on the whole, he would have staged units as a precautionary measure. When Exposure 4 asked for additional units, Command already would have had them staged and ready to go! Command has control over the entire scene while staying at the command post. He should let someone else do the walking.

THE RELATIONSHIP BETWEEN COMMAND AND EXPOSURE

There needs to be trust between Exposure and Command. As Command, I want an experienced officer assigned as Exposure. In many cases, I will assign an exposure officer and then "turn my back" to the exposure. I may never even see the exposed structure until after the fire has been extinguished. I need to know that I have an officer who will ask for only what is needed to handle the situation, and not make a third alarm out of a curtain that has caught fire in an open window on the second floor. Having said that, I also want to know that the officer I have assigned to Exposure will not be afraid to get all the help he needs to confine the fire to the smallest area possible.

Exposure officers need to know a variety of firefighting strategies and tactics. They, like Command, also must know which other officers are on the scene and their specific personality strengths and weaknesses. Which officers are alarmists? Which are the "lone rangers" who want to handle everything within sight by themselves at the expense of their crews? That is the reason Command should choose his exposure officer wisely. Experience and trust are the keys.

Figure 15-7. Exposure protection at a large incident. Notice the direction of the aerial stream. Sometimes you hope you can keep the fire to the block of origin.

RESPONSIBILITIES OF EXPOSURE

Exposure must keep in mind three responsibilities as he moves throughout the structure.

- *The safety of his crew.* The member on the exterior hoseline will need to be reminded of his location and the effects of radiant heat. The officer does not always remain with his crew in this sector. If the exposure becomes significant, one crew member should remain to operate the exposure line while the exposure officer and a crew member go to check the inside of the exposed building. The officer must choose the best-qualified firefighter to remain outside with the hoseline.

- *Protect the assigned exposure.* This will require not only the application of a hose stream but also gaining entry into the exposed area to ensure that the fire is kept out.

- *Keep Command informed.* Command must be updated periodically on the condition of the assigned exposure or area.

Endnotes

1. These horizontal and vertical movements are influenced by the wind, even in an enclosed building.

2. The major spread of fire from floor to floor via exterior windows is called "autoexposure" or "lapping."

3. An often misused term is "occupied/unoccupied." Both words indicate that the home (or structure) is inhabited. Occupied means the structure is being used for habitation, whether the occupants are home at the time or not. Occupied is the opposite of vacant. Unoccupied (which often is substituted for vacant) indicates the occupants (who do live there) are not home at the present time.

4. Lee side refers to the side to which the wind is moving. Windward side is the side from which the wind is coming.

REVIEW QUESTIONS

1. What are the four means of heat transfer?

2. List the two types of exposures.

3. What are the two types of interior exposures?

4. What factors influence the priority of exposure protection?

5. How are interior horizontal exposures protected?

6. How are interior vertical exposures protected?

7. How are exterior exposures protected?

8. What are the two most frequent mistakes the author cites with regard to protecting exterior exposures?

9. Which method of heat transfer is most responsible for interior exposure problems?

10. Which method of heat transfer is most responsible for exterior exposure problems?

DISCUSSION QUESTIONS

1. How do the factors that influence the priority of exposure protection relate to life safety and economic value? Should they be concerns? Should they be the *only* concerns?

2. Discuss fires to which you responded where exposures were a problem. Did your incident commander assign a separate officer to handle each specific exposure, as the author suggests, or were the exposures all under the immediate direction of Command? What were the outcomes of these incidents?

3. The author lists two problems he observes crews making related to exposure protection. Can you cite other "errors" made in protecting exposures?

16

The Mission of Extension

KEY CONCEPTS

✓ Autoexposure

✓ Extension

*T*he mission of Extension is to check the areas above, surrounding, and below the main body of fire for extension and to report the findings back to Command. For the most part, it is Attack's responsibility to check for extension. If Command designates one attack sector (not differentiating between geographical areas), then Attack is responsible for the fire in the entire structure or area. Once all visible fire is knocked down, the attack sector should notify Command. Command should begin to see noticeable differences in fire and smoke conditions as soon as the main body of fire has been extinguished. The color of the smoke will be lighter, the force with which it is exiting the structure should lessen, there should be less glow, and visibility should begin to improve throughout the structure. The next step is to check for extension.

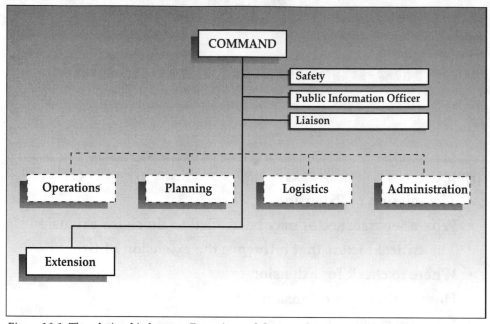

Figure 16-1. The relationship between Extension and Command.

This doesn't mean that the fire is out. We have all heard of units leaving the scene too soon. Rekindles occur. At this writing, the insurance industry is beginning to file lawsuits against departments that have had rekindles occur (and they're winning)! This puts a whole new slant on checking for extension and overhaul.

WHEN EXTENSION SHOULD BE A SEPARATE SECTOR

As already stated, we normally do not need to assign a specific and separate company to check for fire extension. More than 90 percent of our fires are handled by one line, and Attack can check for extension after the fire has been knocked down but prior to overhauling the area. However, there will be fires where Attack has his hands full just putting out fire. At some of these fires, Command may want an earlier accounting of the fire and the direction in which it is traveling. At these times, Command may choose to assign an extension sector.

Note: For the sake of clarity, I wish to point out that normally the attack sector is responsible for checking for extension. This usually is done as a deliberate effort immediately after the main body of fire has been knocked down or during the overhaul process. However, there may be times when fire conditions necessitate that Command receive a quicker accounting of the fire's traveling pattern. In those instances, Command may opt to assign an extension sector.

Instances in which Command may want a more timely accounting of the extension of fire would be the following:

Figure 16-2 (top left). An occupied wood-frame house. Figure 16-3 (top right). A high-rise structure. Figure 16-4 (bottom left). A garden apartment with a common attic. Figure 16-5 (bottom right). Most times, the strip mall, a relatively new concept in retail and construction, has lightweight truss roofs and common attics. (Photos by author.)

- *Occupied wood-frame structure other than a single-family dwelling.* Apartments, such as garden type, are prone to vertical spread by their very design. Fire can quickly enter construction voids and plumbing chases, which are in line from floor to floor, and travel to upper units and common attics. Additionally, older balloon-frame, multifamily structures are extremely prone to fire spread in *all* directions.
- *High-rise structures.* These buildings are prone to fire spread by interior chases, stairwells, and shafts, by autoexposure—i.e., fire spreading from floor to floor through outside fire extension via windows, also called "lapping." In a working high-rise fire, it is essential that companies quickly get above the fire to check for vertical spread.
- *Structures that share the same cockloft or attic areas.* Row houses, brownstones, and commercial two- and three-story ordinary-constructed buildings in older districts are especially susceptible to quick vertical spread.
- *Strip malls.* A relatively new phenomenon in commercial America, these malls—known as the "taxpayer" of the '90s—are constructed to be put up quickly and cheaply. Almost all of them have lightweight steel-bar joist or wood-truss roof assemblies, which are very susceptible to early collapse. A common practice is to remove ceiling tiles in storage areas in these occupanices, which facilitates fire spread. Once fire enters the truss loft area, horizontal spread is rapid. Many fire stops or fire walls are insufficient or have otherwise been compromised.

In the previous instances, and others, it is crucial that Command have someone immediately check to determine whether fire has spread beyond the area of origin.

CRITICAL FACTORS INFLUENCING THE EXTENSION OF FIRE

Heat rises as a result of convective currents. These currents rise until they meet an obstruction. Once they meet an obstruction, they travel horizontally until they can rise again. This fact dictates that the spread of fire is generally higher than the point of origin. Radiant heat (and, to a lesser degree, direct-flame impingement) spreads fire horizontally from the point of origin. For the most part, however, the most significant spread of fire is upward.

Construction type and method to a large extent influence spread and thus extension of fire. Concrete construction can be made to be virtually spread-free—the contents will be susceptible to fire, but any spread from compartment to compartment can be stopped. With rated and properly installed and maintained doors, concrete structures tend to become very compartmentalized. That's the good news: There is little spread. The bad news is that there is no place for the heat produced in fires in these structure types to go. Fires in these structures tend to be extremely hot and, consequently, difficult to approach. Wood-frame structures, especially balloon frame, are extremely prone to spread and thus extension of fire. Contrary to common belief, the spread of fire in balloon-frame structures is fairly predictable: It can spread everywhere! My rule: If fire enters the structural framing system, open everywhere you believe fire could be. Platform-frame construction is less prone to spread than balloon construction, but the structure still is made primarily of a combustible material, and, as such, spread can be anticipated.

The last critical factor I want to discuss concerning the spread of fires, otherwise called "extension," is that of multiple fires. It is predicted that 40 percent[1] of the fires in urban areas of the United States are incendiary. Normally (although I have never seen any statistics to substantiate this), when a fire is deliberately set, the arsonist will set several fires—usually in closets or natural openings in the structure such as stairways and around ductwork. This fact makes it wise to send crews around the structure to look for additional fires. These fires may also be "trailered"[2]—in the means of egress of the arsonist—another reason to check for extension.

WHERE TO CHECK FOR EXTENSION

Three places need to be checked when looking for fire extension. In order of importance, they are

- Above the fire(s). As stated earlier, heat rises. Convective currents will carry heat upward, toward the ceiling, while pulling oxygen and cool air down to the lower portions of the fire room. This rising heat is the reason that areas directly over the fire or over its direction of travel should be checked first. (The wind and other natural factors will influence interior spread and thus the direction of travel.) Heat rises not only in open air but also in construction voids and channels. Holes in ceilings due to lightning or other fixtures will allow heat and fire to pass to upper regions of the structure. If fire enters a stud-wall assembly, construction features or worker error may permit heat to pass to upper areas.

- *Around the fire(s).* Next, check the wall assemblies in the room around the fire area. Normally, attack crews should handle the wall surfaces in the fire room.[3] Usually fire will spread from room to room by convective currents inside wall assemblies, radiant heat, conduction, and direct flame impingement. Fire can enter wall assemblies and ignite lath on the opposite side. Exterior wall surfaces must also be checked. No matter what the fire route, horizontal spread must be considered.

- *Below the fire.* Fire can drop down into the floor joist system and travel horizontally. This is true of platform as well as balloon construction. I have seen fire drop down stairways, igniting combustibles near the ceiling level and traveling horizontally in an attempt to find another way up. Fire and hot embers will drop down laundry chutes and heating ducts. Fire, heat, and embers will travel inside exterior wall assemblies in balloon-constructed frame buildings.

When looking for extension above the fire, check the area immediately above the room of origin first. In two-story homes, use the stairways, if possible, for the quickest access. Areas at the tops of stairs should also be checked. This is the general avenue of travel from floor to floor. Remember that heat traveling convectively can reach 1,400° and higher— higher than the ignition temperature of most common combustibles. (Areas in the vertical path of travel can be subjected to extreme heat.) Check the exterior walls above the fire room to make sure that fire did not climb the exterior walls. Finally, the check for fire extension should proceed in the direction of the wind. If fire is pushed by the wind, it will move in that direction.

When checking for horizontal spread around the fire room, consider three factors:

Figure 16-6. A fire crew checking for the extension of fire above the fire room in a vacant apartment. Notice the tools in the photo.

- *Location of the original fire.* Fires started by whatever means in the center of a structure will require more horizontal checking than fires originating in areas bordering exterior walls. Mathematically, there are just more areas to check.
- *Wind direction and speed.* These factors significantly influence interior fire travel and thus areas of possible horizontal extension. Give priority to rooms adjoining and downwind of the fire room. The greater the wind speed, the greater the chance of horizontal spread.
- *Construction features.* These elements influence construction type and thus extension. Drywall[4] on steel studs is certainly less prone to extension than lath-and-plaster walls. Old homes may have wall registers that allow furnace heat to pass from one duct to be shared between rooms, allowing for horizontal spread.

The same rules basically hold true for checking for extension above the fire as for areas below the fire. Start in areas directly below the room of origin and move into areas in the direction in which the wind is traveling. Check open passageways and shaft areas like laundry chutes and ductwork that come off or are contiguous with the chutes or ductwork. A newer aspect of wood-frame construction is that of truss floor assemblies. The good news is that this type of floor-joist system is in newer homes and structures that are of platform construction, which is less susceptible to spread and should be up to code. The bad news is that if, due to worker error or neglect, fire is allowed to pass into this truss floor area, fire loads and oxygen content are great. All that would be needed is heat.

CHECKING FOR EXTENSION

Specific tools are needed to check for extension. My first tools of choice are the firefighters' senses.

- *Sight.* Look! Normally, an experienced firefighter will have a gut feeling as to whether the fire has spread just by entering an area. Look for fire, glowing embers, or smoke coming from wall openings, behind floor moldings, and moving up the face of a wall. Smoke will naturally be present in upper areas of a structure being assaulted by fire. The force behind the smoke and the smoke's color are good indicators of potential spread. The more force and the darker the smoke, the more problems.
- *Sound.* Listen for crackling noises and the hiss of air being forced through tight spaces.
- *Feel.* Check the heat directly over wall channels in the attic area of balloon-frame structures. In platform structures, check for heat in areas around the locations at which the double plate and rafters meet. Utility holes may have been drilled through the double plate to allow for the passage of cable, wire, and the like. Feel exposed wall surfaces. The old rule of thumb is, If you can put your bare hand on the wall long enough to say, "It's not hot, Chief," there is no fire in the wall. Remember that this

rule of thumb probably originated prior to the use of drywall. This rule works best on lath-and-plaster assemblies. Drywall's ability to collect and retain heat (an endothermic material) makes it extremely hot to the touch. However, except for rare exceptions (normally caused by improper taping or other installer error), fire has a hard time entering drywall assemblies.

Also, feel floors for excess heat. Before the advent of bunker pants, it was not uncommon to have your knees warm up considerably while crawling over a floor assembly that had fire inside it.

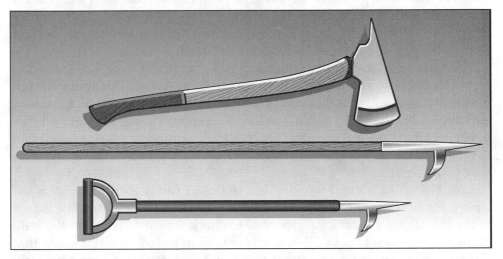

Figure 16-7. The fire ax, pike pole, and closet hook are the tools of choice when checking for fire extension.

Hand tools of choice are fire axes, pike poles, and closet hooks. An economic way to check for extension above and around the fire is to carefully remove the floor molding on the fire floor and the area (room) directly above the fire. (Some of this molding is old and hard to replace. If the extent of fire is questionable, care should be taken.) After a small section of the molding has been carefully removed, a hole can be punched where the molding was. If there is no extension of fire, the occupant can easily put the molding back with no visible side effects.

If in doubt, open up. Placing a small hole in the ceiling or wall is certainly better than coming back 20 minutes after leaving the scene and finding the house rolling again.

Remember that conserving property is one of the four priorities of Command. The aspect of checking for extension plays a great role in property conservation. It would be easy (other than for being physically tiring) to open up an entire structure every time we have a fire. That, however, would be a waste of effort and owner/insurance company money. First, check for extension to ensure that the fire has been confined to the area of original involvement. Get ahead of it, around it, and below it to cut off its spread and reduce dollar loss. Second, open up with caution and prudence. If in doubt, open up a section for inspection, but only if you believe the fire could have entered there. The less damage done while checking for extension, the lower the dollar loss will be.

WHEN TO ASSIGN EXTENSION

As stated earlier, Attack usually checks for extension. If the magnitude of fire is great and Attack's hands are going to be full for quite a while,[5] Command should assign another crew to extension. If the structure is vacant, Command probably can assign a company to extension and, after it gives the benchmark,[6] reassign the company to search. (These assignments could be switched if the situation dictates and search is higher on Command's to-do list than checking for extension.)

At multifamily residential structures and apartment fires, the extension sector probably would have a high place on the to-do list. This is due to construction features and the potential for problems with civilian safety. Although I normally do not like to have a company assigned to two functional sectors at once, a working fire in a multifamily structure may dictate that a company be assigned to search and to check for obvious extension as they go along.

A final word: At normal working house fires, I rarely assign an extension officer, for two reasons. First, most fires are handled by one line. The extent of the fire allows the attack sector time to check for extension after the bulk of the fire has been knocked down. I have faith in my officers and know that if they get inside and suspect that they can't effectively and expeditiously check for extension, they'll tell me. Second, like most of you, I just don't have the personnel. The "Big Four," discussed earlier, tend to take up my personnel allotment. To special-call for an additional engine to do what Attack probably can do before that engine gets there doesn't make sense.

SHOULD EXTENSION HAVE A HOSELINE?

Whether Extension should have a hoseline is for Command and the extension officer to determine. If fire conditions indicate that fire likely entered other areas in significant proportions, it would be prudent to give a line to the extension sector. The primary responsibility of the sector officer is the safety of his crew.

But, let's think about this. If I believe the fire has extended to other portions of the structure and Attack cannot handle all of the fire, then why not just assign the company as another attack sector? Instead of ...

Command to Engine 6: You're Extension,

try ...

Command to Engine 6: You're Division 2 Attack.

Now the focus and expectations are clear. Additionally, you're telling the initial attack officer you are aware that there is or may be fire in another area and you don't want the original attack sector to worry about it.

I normally take the above route over worrying whether Extension needs a line. If I feel that chances are that the fire has not extended from one area to another—but I want to be sure—I'll assign an extension sector, and I do not expect it to take in a line. I should have a backup crew in place. Backup should be made aware that I have a company inside checking for extension and that he may have another crew to worry about and may need to reposition the backup line. If it is likely that the fire has extended, I'll just assign another attack sector.

If the extension officer gets to the area(s) being checked and feels uncomfortable about going farther, he should inform Command. Once notified, Command has three options:

- He can tell the extension officer to come out and assign another company to attack in that area.
- He can tell the extension officer to come out and get a line and that he will change his designation from Extension to Division ___ Attack.
- He can have the extension crew come out and have the original Attack move to extinguish that fire.

THE RELATIONSHIP BETWEEN COMMAND AND EXTENSION

Command needs to have faith in the extension officer. Extension's judgment may have a great bearing on the outcome of the fire. An experienced officer with an understanding of construction features is ideal for the assignment of extension officer. There must be communication between Command, Attack, and Extension concerning probable and potential problem areas. Command needs to determine the construction type if it is not known. This information needs to be relayed to Attack and Extension as soon as possible. Conversely, Extension can't check out areas and then leave Command or Attack guessing. Once an area has been checked, the outcome must be relayed over the radio immediately so subsequent moves can be made. Up-to-date communication between these three sectors is vital to the quick, efficient extinguishment of the fire.

THE RESPONSIBILITIES OF EXTENSION

The extension sector officer has the following responsibilities:

- The safety of his crew. Extension will not have a hoseline. If the officer assigned to extension believes he may be cut off by fire or that he will otherwise be placing his crew in jeopardy, he should stop moving toward areas where fire could be spreading. He should inform Command, who will (1) have the crew get a line and reassign them to attack in that area, (2) assign another company to attack in the area in question, or (3) have Backup move into a better position while the extension crew moves in to check the area.

- Check the area above, around, and below the fire for extension.
- Keep Command informed. Command created an extension sector because he was concerned about the possible spread of fire within the structure. Before moving from between, above, around, and below the fire, Extension must inform Command of his findings. Command will confirm Extension's findings or change his "direction" or assignment. But Command must be informed first.

Endnotes

1. DeHaan, John D. *Kirks Fire Investigation,* Third Edition. 1991. Brady/Prentice Hall Publications, 2.

2. Trailers are long, narrow pools of liquid fuel or pieces of paper soaked in fuel. They act as a fuse for arson fires. Arsonists will construct a fire area and then pour or arrange a trail of fuel in the direction of the exit. The trailer is ignited just before exiting the building or fire area.

3. Discussion of normal overhaul practices is covered in the next chapter. Overhaul and checking for extension are similar but two distinct phases (sectors) of a structure fire.

4. Drywall (gypsum) is extremely endothermic. Steel studs will absorb some heat from the drywall; the heat will spread by conduction but normally not to the extent that ignition of combustibles will occur.

5. "Quite a while" is a relative term. If the fire has gained a sufficient hold on an entire room and Command estimates that Attack will not be able to check above the fire for at least five minutes, then he'd better get someone else started up above the fire. This is contingent on having available units staged at the scene or close by.

6. Benchmarks are specific announcements given by sector officers. They are discussed in Chapter 19.

REVIEW QUESTIONS

1. What is the mission of Extension?

2. Who usually checks for fire extension?

3. List in order of importance the three places that must be checked for fire extension.

4. In the author's opinion, what are the tools of choice when determining where to look for fire extension?

5. What are the hand tools of choice?

6. What three critical factors influence fire extension?

7. If extensive fire may be in the areas in question, which of the following should be done: (a) assign an extension sector, or (b) assign an attack sector?

8. Who in addition to the extension officer should be looking out for the safety of the extension crew?

9. What are the responsibilities of Extension?

10. If the extension officer feels uncomfortable about the area that needs to be checked, what are Command's options after having been notified of this concern?

DISCUSSION QUESTIONS

1. Compare and contrast the attack and extension sectors.

2. Discuss extension crews' going above a fire without a hoseline.

3. Discuss knowledge of building construction as it relates to checking for fire extension.

The Mission of Overhaul and Salvage

TOPICS

- The mission of Overhaul
- Tasks Overhaul must complete
- When overhaul should be assigned
- Where and how to overhaul a structure fire
- Common wall types found in residential structures and how they relate to overhaul
- Options and cautions involving trash accumulated as a result of overhaul
- Determining the origin of the fire
- The difference between overhaul and checking for extension
- The responsibilities of Overhaul
- The mission of Salvage
- The components of salvage
- When to assign salvage
- Where and how to conduct salvage operations
- The responsibilities of Salvage

The mission of Overhaul is to ensure that the fire is completely out and to pinpoint the area of origin. I must admit that I normally do not think of overhaul at most single-family structure fires. In most departments, there seems to be a smooth transition from attack to extension to overhaul. Normally, these three individual (and I must add "distinct") functional sectors are carried out by Attack during our bread-and-butter fires. (As I write this, I am convincing myself that this should be the exception, not the norm, at a working fire.) These functions should be distinctly assigned by Command. They can, and in most circumstances are, given to the same

KEY CONCEPTS

✓ Area of origin
✓ Endothermic
✓ Nozzle discipline
✓ Overhaul
✓ Point of origin
✓ Salvage
✓ Salvage covers
✓ "V" patterns

company *after the company has given the appropriate benchmark.* I distinguish "working fire" because at a small, part-room fire, such as one in a closet or mattress, the one or two units that end up handling the fire—normally an engine and a truck company—handle all of these functional sectors and more.

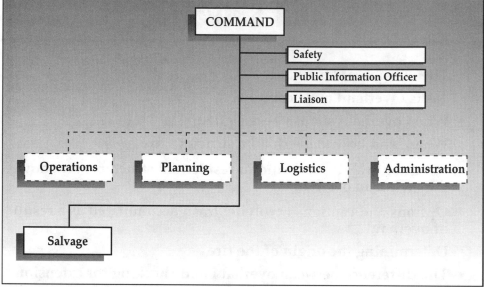

Figure 17-1. The relationship between Overhaul and Command.

SCENARIO

Pretend you are a chief officer responding to a working fire in the west end of town. The first unit on the scene reports fire showing from the second floor of a two-story occupied home and takes command.

 Engine 7 at 1945 Vermont: We have fire showing on the second floor of an occupied home. Engine 7 will be Vermont Command.

Command assigns the remainder of his crew to attack and has the second-in engine lay a supply line in to them and then assigns them backup. Command assigns the truck topside ventilation and the rescue squad search.

 Command to Engine 7: You're Attack. Engine 5, lay into Engine 7 and then take Backup. Truck 7, you're Ventilation. I need you on the roof. Squad 7, you're Search.

You, as chief, listen while responding. You believe that all the initial bases are being covered. "I've trained them well," you say to yourself. You arrive on the scene, get a briefing from Command, and then assume command and send the officer of Engine 7 to join his crew and take over as attack officer.

Battlion 1 is now Vermont Command.

After the fire has been knocked down and the building vented, you go inside to determine the origin and cause. You meet Attack on the first floor and ask him where he believes the fire started. He responds, "Looks like the bedroom in Sector A." You go up, and it looks as if the fire started in an area where a curling iron was plugged into the wall outlet. There is beading about one inch beyond the wall plug on the curling iron cord. You go outside and talk to the occupant. She says she was using the curling iron and that she has had problems with the iron's not heating up all the time and that she was going to buy another one today. She explains that the phone rang as she was doing her hair and that the detector went off while she was talking on the phone. She adds that she ran to her room and found the curtains near the wall plug burning.

You now believe you have your cause and origin. You tell Attack you're going to leave the scene and that he will be Command. You give him a few last-minute instructions concerning the electric company and where to put the debris. You leave.

Battalion 1 will be returning to service on Channel 1. Engine 7 will now be Vermont Command.

Dispatch: Okay, Battalion 1, 1347.

At 1358, Dispatch notifies you on Channel 1 that the incident on Vermont is complete.

At 1438, you are dispatched to a fire at the same home on Vermont. Engine 7 reports on the scene that there is heavy smoke on Division 2 and takes command.

After this fire has been extinguished, you determine that the fire had extended up an exterior wall on Side 2 of the structure and heated up some cellulose insulation in the attic. The PPV fan may have helped push the fire up the wall. You question the officer on Engine 7, who said that when you left, he had his crew clean up the bedroom with Engine 5. He said he thought Engine 5 would check the walls and the attic. Truck 7 officer said that they'd had smoke but not much heat coming from the vent hole they'd put in the roof. He added that he thought someone else had checked that attic. Squad 7 said they left right after you did. Engine 5 officer said that after they cleaned up the burned "stuff" in the room, they helped pick up hose while the officer on Engine 7 went back inside to take one last look around.

Who is responsible here? Is this a rekindle that could have been avoided? Does any of this sound familiar? Officers have to assume certain things at incidents. This can lead to problems if communication and assignments are not specific. As stated throughout this text, Command is ultimately responsible for the outcome of the incident. In this case, everyone was a little guilty.

Command should have assigned Attack to overhaul as soon as the fire was knocked down and the benchmark given:

Attack to Command: We have the fire knocked down.
Command: Okay. Your new designation is Overhaul.

Someone would have been responsible for the overhaul process, and there would have been no assuming or guessing.

Hopefully, the officer of Engine 7, who was left in charge after the chief left, knew better than to have left the scene of a fire in the second floor of a two-story frame structure without checking the attic. He, as the chief, should have asked or should have been specific in making the assignments. Finally, he should have ensured that the attic was checked prior to leaving, or he should have checked it himself.

The officer on Engine 5, who was left inside to clean up the room, is not without guilt at this incident. After the chief left the scene, the officer of Engine 7, who was Command, went outside and had his crew and the crew of Engine 5, including the officer of Engine 5, clean up the fire room. The cause had been determined, so what needed to be done could have been done.

A prudent officer would have removed the wall plug and checked the area behind the plug. (This area is prone to dust accumulation and even cobwebs, both of which burn at very low temperatures.) Additionally, he would have felt the wall surface in the area of ignition (as well as the walls in the entire room) for heat and then again later to see if the walls were cooling down. He also should have sent someone up to the attic just to "watch" over the area of ignition for a while, looking for wisps of smoke.

Because no one was actually assigned to overhaul, responsibility for this incident has to fall back on the chief officer and the officer of Engine 7. Both neglected to assign a sector that should have been assigned.

WHAT OVERHAUL ENTAILS

As stated in the mission statement, Overhaul has two tasks to complete:
- to ensure that the fire is completely out, and
- to pinpoint the area of origin.

Overhaul is one of the few functional sectors that can be assigned at a fire incident that has this dual role. Other sectors, to be sure, have many things to do to complete their mission, but they have only one basic task—to search, put out the fire, vent the building, and so on. Overhaul has two specific (and basically different) tasks. The order in which Overhaul completes these tasks may appear to be a "no brainer"; however, they need to be done in the reverse order of that which might be chosen by an "untrained" person. You would think that it is more important to put the fire out than to determine the point of origin. However, the bigger picture needs to be looked at here.

Figure 17-2. A firefighter overhauling the fire area.

Fire cause and determination are directly related to the area of origin. Once the area of origin has been determined, the actual cause of the fire must be accurately pinpointed through a thorough and systematic investigation by a trained individual. If firefighters are allowed to overhaul (and overhaul completely), they may (and chances are, they will) destroy all evidence needed to pinpoint the cause of the fire. If a crime has been committed (arson) and evidence has been taken from areas related to the point of origin due to the overhaul process, then conviction(s) may be difficult, if not imposible. To accurately determine the cause, the point of origin must not be disturbed during the overhaul process. This means that the point of origin must be determined before beginning overhaul. Once the area is isolated and protected, an investigator can survey the area, look for specific clues, take samples, and then document the area as completely as possible using photos or video. After these processes have been completed, the area can then be completely overhauled.

Once the area of origin has been pinpointed *and* the scene investigation process completed, all fire-damaged property and probable (not only evident) hot spots must be extinguished or removed for final extinguishment.

WHEN TO ASSIGN OVERHAUL

Let's start out by saying that if overhaul is not assigned, the scenario provided earlier in this chapter may end up being a reality. The beauty of the incident management system is that roles and responsibilities are defined. If no assignments are made, the responsibility falls back to the individual in charge—Command. If assignments are made, then the sector officer assigned should be absolutely, positively sure of his role and responsibility. If mission statements are used, there should be no misunderstanding. If Command assigns overhaul and a rekindle occurs, there is no question as to who should have located and extinguished any and all potential hot spots. (The incident management system is not—and should not be—a vehicle for placing blame. It is a tool that defines "who is doing what—where" and what every respondent's role and area of responsibility are *in that particular incident.*[1])

Overhaul should be assigned as soon as the main body of fire has been knocked down. Delays in assigning an overhaul sector can allow hot spots to flare up and, if undetected, can allow hidden fire to gain in size and strength.

WHO SHOULD BE ASSIGNED TO OVERHAUL?

For normal fires, this is a natural. Attack is there and should have all the tools necessary to commence overhaul. Once the main body of fire has been knocked down (and after the appropriate benchmark has been given), Command should reassign Attack to Overhaul:

Figure 17-3. Fire crews overhauling a fire area. When overhauling a mattress that has been involved in fire, it's best to get it outside. Watch for flare-ups.

Attack to Command: We have the main body of fire knocked down. I don't think it went anywhere.

Command: Okay. Your new designation is Overhaul.

Engine 6: Okay on Overhaul.

The one exception to this is when staffing is low and

fires are or may be in other areas of the structure. If Command assigns one attack sector and more than one area may be involved, Attack should move to the next logical area to seek out and extinguish fire.

WHERE TO OVERHAUL

After the area of origin has been determined, overhaul should be conducted above, around, and below the fire (following the same rules as when checking for extension). If the fire was small with little vertical or horizontal spread, the areas above the fire can be quickly checked and the overhaul crew can move on to more localized areas around the fire. The most problematic areas are stuffed furniture, wall assemblies, closets, and certain floor coverings in areas in close proximity to the fire. When the fire is confined to one room, the entire room should be checked. Look for soot and signs of heat (char) on ceiling surfaces. If decorative wood surfaces (moldings and the like) have very little charring, the wall plates can be removed and checked for signs of excessive heat.

HOW TO OVERHAUL

The overhaul process should be systematic. The first step is to determine the areas that can immediately be overhauled and the area that cannot be overhauled initially, specifically the area of origin. Start in an area remote from any visible heat damage and work back toward the fire area. Begin by looking for char on door frames and moldings. The areas or sides that have the most char and heat damage are the areas *from which the fire came*. This should lead you to the room of origin. Once the room of origin has been determined, notify Command and seek permission to commence overhaul in other areas. If only one room is involved, you may have to wait for a fire investigator to arrive on the scene (if the chief officer is not responsible for determination). Either way, once the area of origin has been determined and the on-scene investigation completed, the second phase of overhaul can begin.

When overhauling, remove all materials that are or have been smoldering to the outside of the structure. Upholstered furniture should always be suspect. If it has burn marks on it, it should be taken outside and overhauled there. If only cushions are affected, they can be taken outside. Hot spots in upholstered furniture can smolder for hours. If there is any doubt, take it out! Open dresser drawers, and check the condition of the clothing inside. If the dresser is charred but extinguished (no longer burning) and the clothing inside is not affected, the dresser can be left inside or moved to another room.[2]

Walls must be checked for hidden heat. You will need to know what the walls are constructed of to effectively determine if fire is behind the wall without opening it up. *Note*: Heat detectors are now available. These battery-operated handheld detectors can be an aid in the overhaul process. You must practice with them to become proficient in their use. The age of the structure and the general construction type will help determine the type of wall system. If the fire has entered the structure in the area in which you are going to overhaul or in other areas, then you may already have an opening and an idea of the type of wall system. Basically, four types of wall systems are used in single- and multifamily (up to four units) residential structures:

Figure 17-4. An officer assigned to overhaul determining the area of origin. Notice the "V" pattern on the wall behind the sofa.

- *Ordinary.* This system generally is used for exterior wall systems only. The walls may be solid or veneer. If the wall system is of solid ordinary construction, extensive overhaul normally is not required. The brick or block will hold heat; but little, if any, fire will pass through if the wall system is intact. If the wall system is of ordinary brick veneer, it may be necessary to determine what is behind the wall surface and what the brick is laid up against. I have never seen or heard of an interior brick veneer wall with the brick on the inside. It normally is on the outside of the structure. The interior side of the brick veneer wall often will be one of the three wall systems listed below.

- *Lath and plaster.* This type of wall system exists in older homes. I would guess they stopped using it in new construction during the 1950s. Lath-and-plaster walls are constructed by nailing small strips of wooden lath onto wooden studs. Plaster then is spread over the lath and either painted or wallpapered over for a final interior finish. I was taught to find out whether fire had entered a lath-and-plaster wall system by touching it. The old axiom was: If you could put your hand on the wall long enough to say "It's not hot, Chief," there wasn't any fire behind the wall. I have used this "scientific" system many times, and it has worked every time.

Once fire gets behind a lath-and-plaster wall, everything you need for a good fire is there—the heat, the fuel (from the wooden lath), and the oxygen (normally, the construction of these houses permits air to move within the structural framing). If you *think* there might be fire in a lath-and-plaster wall system, you can feel the wall and then wait a little while. Check above in the attic (any smoke will rise, and you can detect smoke

Figure 17-5. A wall assembly made of ordinary construction. (Photo by author.)

and heat in the wall channels in the attic). Feel the wall again for an increase or decrease in heat. Have someone check the attic again. If still in doubt, open it up.

- *Drywall on wood studs.* This wall system replaced the lath-and-plaster wall systems in the 1950s. Drywall (also called "gypsum board,"

Figure 17-6. An officer overhauling a wall assembly constructed of lath and plaster (left of photo).

Figure 17-7. A wall system constructed of drywall on wood stud. (Photo by author.)

Sheetrock®," and "plasterboard," among other names) is made of gypsum, an endothermic material. As such, it absorbs heat in great quantities. The old axiom "It's not hot, Chief" won't work on a drywall assembly because of the amount of heat absorbed. The bad news is that these walls get very hot when involved in fire. The total wall system is only as good as the drywall hanger and drywall finisher. If not hung and taped correctly, part or all of the total wall system can be lost. If you feel you want to open a drywall system to check for fire, punch a hole. The higher up the wall, the better. Why higher up? Heat rises. It costs the same to patch a hole that is high on a wall as it does one that is low. Why not check the area where most of the heat will be concentrated?

- *Drywall on steel studs.* This is a newer form of construction. It generally is found in commerical occupanices or in remodeled sections of existing homes. These wall systems should be overhauled in the same way as a drywall on a wood stud wall system, but note the following. First, there is the potential for an electrical shock. These studs can become excellent conductors of electricity. Second, the possibility—even though slight—exists that the metal stud can conduct heat to other parts of the structure.

ACCUMULATION OF DEBRIS

One of the concerns of Overhaul and Command is where to put the "stuff" shoveled out of a structure—usually materials such as lath and plaster, furniture, floor and ceiling moldings, floor coverings, and so on. At certain fires, you can end up with quite a pile of damaged debris. These piles of debris are two things:

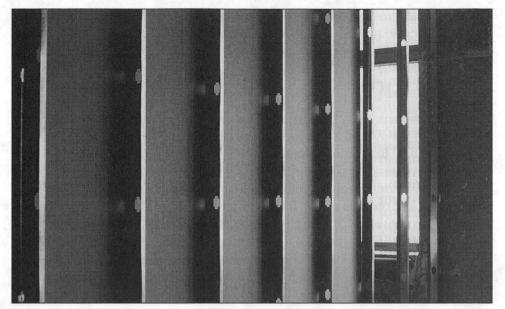

Figure 17-8. A wall system constructed of drywall on steel studs. (Photo by author.)

- *Unsightly*. If at all possible, Overhaul should attempt to place debris in an area where it will be out of view. If there is access to the rear of the build-ing via an alley, the backyard would be an ideal storage area. Along-side homes in drive-ways is another location that may keep neighbors and the owner happy.

- *Dangerous*. Flare-ups of removed material are quite common. The de-bris must be wet down for a long time. Usually this is done after the fire has been extinguished— during the overhaul/ salvage/pickup phase of the fire. The engine driver will pull a boost-er line off the rig and hose down the pile. To save the embarrass-ment of having to respond to a second fire at the same structure,

Figure 17-9. Notice that accumulation of debris became a problem early in this fire.

the smoldering rubbish should not be placed too close to the structure. As mentioned in the previous chapter, it is becoming standard practice in some areas to have insurance companies go after departments that are involved in rekindles. In this case, if a pile of hot debris that was not completely extinguished were left too close to a structure, who would be liable?

In some departments, truck companies carry plastic garbage cans to carry out materials shoveled up in fire-damaged rooms. The cans are cheap and sometimes are less labor-intensive than carrying shovels full of debris through several rooms to an appropriate window.

In our community, legislation requires the owner to remove any fire-damaged debris left outside by the fire department within 72 hours of the fire. If the debris is not removed, the city removes it and bills the owner. The city's Neighborhoods Department[3] follows through on this. Chiefs carry a Polaroid® camera with them. They take a picture of the debris and attach it to the fire report, along with a form the owner (if on the scene) has signed. This seems to be working well.

DETERMINING THE AREA OF ORIGIN

If the area of origin is not obvious, it must be determined prior to overhaul *if* evidence relating to the actual cause of the fire (suspicious, incendiary, or accidental) is to be preserved and examined.

It is difficult to get a conviction for arson. Allowing firefighters to overhaul the area of origin prior to the investigation makes the task of the investigator and prosecutor more difficult. The scene must be kept intact as much as possible so that the actual cause of the fire can be determined and all evidence can be collected and documented.

As mentioned earlier in this chapter, the task of the officer assigned to overhaul is to determine the area of origin. In most cases, the area is obvious. Burned cup-

Figure 17-10. An officer attempting to determine the area of origin. Notice the "V" pattern on the wall.

boards over a gas range, a burner left on under a pan of frying meat, and no burn marks below the cooking surface of the range pretty well isolate and pinpoint the area of origin. The corner of a bedroom where the bed had been that has the obvious "V" pattern would indicate that that was the area or room of origin. Both of these scenarios point to the area of origin and *not the cause*. The fire in the bedroom could have been caused by arson, a child playing with matches or a lighter, an electrical short near the bed, careless smoking, or a variety of other things. In the kitchen scenario, the obvious cause might be considered careless cooking or food left unattended. However, this fire may very well have been intentionally set for profit or to obtain sympathy from a loved one or for other reasons. *Isolating and identifying the area of origin is the beginning—but not the end of determining cause.* Once the area of origin has been identified, it is up to Command or a qualified investigator to determine the actual cause.

THE RESPONSIBILITIES OF OVERHAUL

Overhaul is responsible for the following:
- The safety of his crew. Except for the initial attack, more firefighters are injured during the overhaul process than during any other sector (activity). Nails are stepped on and hands are cut. Members take off their SCBAs too soon and experience breathing difficulty and lightheadedness from accumulations of carbon monoxide. The overhaul officer must continually monitor his crew to ensure that the safest possible conditions exist. Lights should be set up. Fans should be running to move the air within the area.
- Locating the area of origin and then completely extinguishing the fire.
- Keeping Command informed.

OVERHAUL VS. EXTENSION

After reading Chapter 16 and this chapter, there may still be some confusion concerning the difference between extension and overhaul. To refresh your memory: Extension's mission is to check the areas above, surrounding and below the main body of fire for extension and report his findings back to Command. This says nothing about extinguishing any discovered fire. It says, "Report the findings back to Command." The assigning of extension to Attack is a natural progression; it fits neatly between attack and overhaul. However, extension usually is assigned only if Command needs a quick accounting of the fire and its travel *prior to the attack sector's knocking down all visible fire*. If the extension sector finds that the fire has extended to other areas, Command must be notified. He will then decide whether
- Attack should move to the area to which the fire has extended and put it out,
- to assign another company to put out the fire extension, or
- to have Extension take a line and extinguish the fire as an attack sector.

Overhaul's task is clear. His mission is *to determine the area of origin and to ensure that the fire is completely out.* There is only one way to ensure that the fire is completely out: Extinguish it. As stated before, for most fires, checking for extension and overhauling normally are assigned to the initial attack company. (Most fires are handled by one line or less.) These three tasks should be assigned separately and after the appropriate benchmark has been given (Attack: "Fire is knocked down"; Extension: "Extension areas checked"; Overhaul: "Overhaul complete"). Extension looks for fire spread. Overhaul ensures that the fire is completely out.

THE MISSION OF SALVAGE

The mission of Salvage is to protect as much material within the structure as possible from the effects of the fire, its by-products, and suppression efforts. Recall the four priorities of Command discussed in Chapter 2: firefighter safety, civilian safety, stop the problem, and conserve property. *Salvage is conserving property.* But, it is more than that. It is an "attitude." To some, it's part of the job only as an afterthought. To some, it's a burden—i.e., if we spread tarps, we'll have to go back later to pick them up; or, maybe it will stop dripping before we leave. Salvage should be conducted irrespective of neighborhood or economic status.

Salvage is directly related to the following functional sectors:
- *Ventilation.* We ventilate to remove the products of combustion from the structure. Removing heat and smoke is directly tied to salvage; it reduces heat and smoke damage to materials within the building.
- *Attack.* The sooner the production of heat and smoke is stopped, the less damage will occur. Properly placing and advancing the attack line also minimizes damage to materials in adjacent areas. Cut the fire off to the smallest area possible, and then push the fire and its by-products back toward the area of involvement. Finally, nozzle discipline helps minimize damage to materials within the structure. Water damage is a key factor in fire loss.
- *Extension.* By rapidly checking above, around, and below the fire area, spread and consequent damage can be cut or stopped.
- *Overhaul.* During the overhaul process, savable items (including parts of the structure itself, such as floor moldings and stained-glass windows) will be accounted for and removed or otherwise protected. Overhaul should only overhaul what was damaged by fire.

Communication, coordination, and understanding among these sectors at a structure fire are musts. Salvage is not a complicated concept. The key is to protect the contents of the structure from the effects of the fire and its by-products, fire control efforts, and the firefighters themselves.

WHAT SALVAGE ENTAILS

Salvage can be broken down into four basic categories:
- *Covering furniture and other "items" with salvage covers.* These covers are available in numerous types and sizes and for various applications

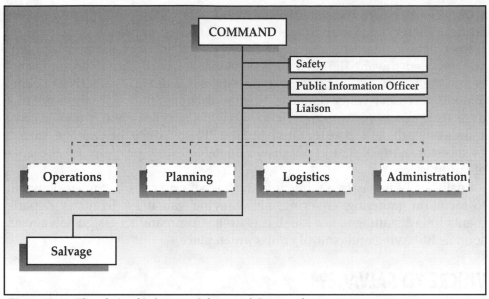

Figure 17-11. The relationship between Salvage and Command.

(throws). These evolutions will not be covered here.

- *Separating areas from the fire and its effects.* Simply closing doors can protect areas from heat and smoke.
- *Protecting valuables.* In my opinion, this entails more than covering items with salvage covers. It also includes quickly removing photos and other memorabilia from walls and taking them outside or to other safe areas or not allowing overhaul to shovel the family Bible out with debris, for example. This approach to salvage probably does as much for our public image as putting out the fire.
- *Reminding sector officers that they should prevent unnecessary damage.* (This can be accomplished in large part through training.)

WHO SHOULD BE ASSIGNED SALVAGE?

Truck companies usually are trained to accomplish salvage. They generally carry salvage covers and rug runners and more squeegees than other units. However, if salvage pops to the top of Command's to-do list, any company could be assigned to this sector; therefore, all companies should be trained in basic salvage operations.

WHEN TO ASSIGN SALVAGE

At vacant houses that have had previous fires, Command may never assign salvage. The contents are gone and whatever is left probably is of little or no value. At an occupied structure, salvage should be assigned as soon as possible—after the "Big Four"[4] have been assigned. Some texts say salvage should begin with the arrival of the first unit. This is a simplistic approach. Although the statement is true, I believe it represents more of an attitude than actions individuals can take. The best approach is to confine, control, and extinguish. Attack the fire from the area of least involvement, and proceed toward the area of involvement. Check above, around, and below the fire for extension, and so on. Attack should always strive to eliminate unnecessary water damage. Nozzle discipline may not only save our lives (by preserving water prior to receiving supplemental water from a source other than the booster tank), but it will also reduce the damage caused by moisture. Attack should not discriminately destroy property for the sake of suppression.

Realistically, it may be a while into the fire before a crew becomes available to focus on gathering, covering, or removing valuables. In many departments today, staffing is not ideal. It is up to Command to assign salvage as soon as lifesaving and control efforts are in place.

WHERE TO SALVAGE?

Salvage operations should take place in areas that are most likely to contain "savable"property and usually should begin in areas away from the fire area. This makes the task a little easiser. There is plenty of room to work away from other working sectors. Check the areas below the fire floor for water dripping—or, sometimes flowing, if the attack crew is not practicing nozzle discipline—from ceiling lights or fan fixtures.

Closing bedroom, adjacent, and closet doors on the fire floor, above the fire, and below the fire will greatly reduce smoke damage.

To determine the proper areas to be overhauled, the structure must be viewed from the outside. If Command does not give Salvage an indication of where salvage should begin, Salvage should determine the locations after consultation with Command. Questions that Salvage should ask Command would include:

- Where is the seat of the fire?
- What divisions (floors) are involved?
- Where are the attack and other working sectors operating now?
- Are occupants on the scene?

If the second floor has major involvement, Salvage will want to concentrate on the first floor and below initially. If the major area of involvement is on the first floor, Salvage should go up to the second floor and try to check or stop smoke damage and then move to the basement to see if any materials there can be protected from smoke and water damage.

HOW TO SALVAGE

Several techniques are involved in the salvage process. Fire companies usually carry plastic or canvas salvage covers, which can be folded, carried, and placed in unique ways to protect objects. Salvage covers, for example, can be made into little cisterns that will collect water dripping from upper floors. I will not cover here what already has been covered in individual drill manuals and other texts covering firefighting basics.

To protect furniture most effectively, move it to areas where the water flow is least, push it together, and cover it. After the furniture has been gathered and covered, look at walls and on shelves for photos and other objects that can be collected and placed under salvage covers or taken outside the building.

If personnel are available, an officer may be assigned to salvage to focus on the areas being overhauled. Overhaul and Salvage generally are at opposite ends of the spectrum when it comes to intent. The intent of Overhaul is basically to tear stuff apart to ensure that the fire is out. The intent of Salvage is to protect property that has not already been destroyed by the fire, its by-products, or firefighters. If an officer whose focus is on preserving as much property as possible is present during the overhaul process, more property will be preserved. Our task, of course, is to put the fire out, but if savable property can be moved or removed prior to the overhaul process, we can put the fire out *and* save property.

RESPONSIBILITIES OF SALVAGE

The responsibilities of Salvage are as follows:
- the safety of his crew,
- protection of savable items from the effects of the fire and working crews, and
- keeping Command informed (if observations relevant to other sectors or the fire as a whole are observed, Command must be notified).

One final word concerning salvage: Several times in my career, I have seen occupants actually cry from happiness even though their home was damaged by fire. They cried because they thought they had lost everything in the fire and then saw us walking out carrying a photo album, a painting, or a family Bible that had been spared from fire (and the overhaul shovel). Chief William Goldfeder of the Mason Deerfield Joint Fire District told me of the satisfaction he and his crew members felt last May when they located and removed a wedding dress for a woman. A save is a save! Such moments are almost as rewarding as pulling a victim from a fire. Most of the personal thank you's I have received from victims at fires have been prompted by our attempts to save their property. This also is our job!

Endnotes
1. Assignments and areas of responsibility change from one incident to another.
2. Remember that the occupants may or may not have fire insurance. Regardless, the concern here is to conserve property. Throwing items outside through windows without regard for their past, current, or future condition is just plain wrong and, in my opinion, the same as destroying someone's property when no fire is involved. Our task is to conserve property, not just throw items out of windows so we can get back to bed sooner.
3. Neighborhoods is a city department established to maintain harmony and uniformity within the community. Department members handle complaints concerning housing, abandoned buildings, rubbish accumulations, and so on.
4. See Chapter 8 for a review of the "Big Four."

REVIEW QUESTIONS

1. What is the mission of Overhaul?

2. What two tasks are to be completed by Overhaul?

3. When should Overhaul be assigned?

4. Where should overhaul be conducted?

5. What is the first step in overhaul?

6. What four types of wall systems generally are found in residential structures? Give some tips for overhauling these assemblies.

7. What is one precaution that should be taken with debris accumulated during the overhaul process?

8. What is the process for determining the origin of the fire?

9. Is determining the *origin* of the fire the same as determining the *cause* of the fire?

10. What is the difference between overhaul and extension?

11. What are the three responsibilities of Overhaul?

12. What is the mission of Salvage?

13. Salvage is directly related to which four functional sectors?

14. Salvage can be broken down into which four basic categories?

15. When should Salvage be assigned?

16. Where should salvage begin?

17. What unit usually carries the equipment required to conduct salvage operations?

18. How does nozzle discipline affect salvage?

19. What is one of the easiest methods for reducing smoke damage to areas remote from the fire area?

20. What are the three responsibilities of Salvage?

DISCUSSION QUESTIONS

1. Look at the scenario at the beginning of the chapter. At this incident, do you think there would have been an easy flow from attack to extension to overhaul? Would that have prevented the rekindle, or are rekindles inevitable?

2. Explain the difference between extension and overhaul.

3. The author points out that salvage is an "attitude." Reflect on your department's salvage procedures. Do you agree or disagree with the author's premise?

18

The Mission of Rapid Intervention Teams (RIT)

TOPICS

- The mission of a rapid intervention team (RIT)

- How NFPA 1500 relates to RITs

- Who should act as a RIT

- Where a RIT should be established

- Tools a RIT should have

- Where a RIT should stage

- The items in a RIT's scene survey

*T*he mission of the rapid intervention team (RIT) is to search for and remove trapped or injured firefighters. The concept of rapid intervention is relatively new to the fire service. I do not know how or when this concept originated. Perhaps it was mirrored from water rescue teams, which as standard practice have an individual(s) suited up and ready to enter the water in case the rescuer gets in trouble. Perhaps it sprang from concepts established by haz-mat technicians, whereby a backup team, suited in the appropriate level of clothing, stands by while the entry team works to mitigate the situation. Wherever the concept came from, I believe it has valid-

KEY CONCEPTS

✓ Emergency traffic

✓ Hauling line

✓ Rapid intervention team

ity and needs to be considered. The City of New York (NY) Fire Department has used rapid intervention teams, called "FAST" (Firefighter Assist and Search Teams), for more than three years now. FAST teams in FDNY are primarily truck companies used at the call of the on-scene chief. As a result of several deaths that occurred in rapid succession about three years ago, FAST teams have become standard procedure in New York City.

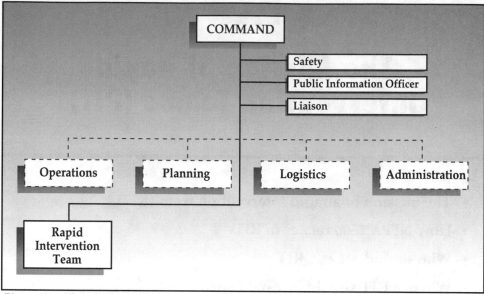

Figure 18-1. The relationship between the rapid intervention team and Command.

NFPA 1500, *Standard on Fire Department Occupational Safety and Health Program*—1992,[1] calls for the establishment of a RIT "at any incident where the need arises." That statement is certainly open to interpretation. The same standard goes on to state that "a RIT is needed whenever fire crews are operating in places or instances that would subject them to immediate danger." Again, this is open to interpretation.

I have discussed this concept with both of the heavy rescue companies assigned to my battalion. They are aware of the concept, and I have used a RIT several times in the recent past. Since the Bricelyn Street fire in Pittsburgh about two years ago, the Pittsburgh Fire Bureau has implemented this concept (called "Go" teams). It is used at every working fire in Pittsburgh.

THE CONCEPT

As we know, Command's task is to focus on the whole incident. A sector officer's task is to tunnel in on his specific assignment. Attack should put the fire out. Search should look for and remove viable victims. Backup should focus on protecting working crews. At some incidents, a backup crew is not enough. The situation is such that there is reason to suspect two things:
- Savable people or things are inside the structure. (If there were no savable victims or things in the structure, there would be no reason to have crews at risk inside.)
- Crews working inside to remove victims or protect these savable things are at a more-than-usual risk of injury or entrapment.

The purpose of the RIT is to have on-scene a specific crew all dressed and ready to go should fire personnel within the structure become trapped, lost, or injured. This crew does not enter the structure at any time during the incident unless it is needed to rescue fire personnel operating therein. It has no other assignment. It is there specifically to find and remove firefighters who for any reason cannot remove themselves from the structure.

THE IMPLICATIONS OF NFPA 1500

NFPA 1500 defines the need for a RIT. It states that a team is necessary "if crews are or could be in immediate danger."[2] The standard specifies that in the early stages of an incident, a RIT "can be either specific and standing by to be deployed if needed or a crew with another assignment that can be redeployed at any time." The standard goes on to state that "as the incident expands in size or complexity," a RIT can be "on scene" or a specific "staged" unit. Although the standard says the RIT can be a crew with another assignment, I do not advocate this. It all goes back to focus. If I assign Search to be the standby RIT, where will the search team's focus be? In addition, the RIT should have specific tools with it, and it has specific tasks to perform.

Of course, certain personnel constraints must be considered. We all have had situations in which the tasks to be accomplished (Command's to-do list) exceeded the number of personnel available to do them. It seems that in today's firefighting world, this is the norm rather than the exception at working incidents. If I make a precautionary second alarm in Toledo, I put the rest of the city at a severe shortage. If another working incident comes in, we are at recall status. To assign a RIT at normal structure fires is a luxury I usually cannot afford. However, like the special tools we firefighters have up our sleeves, it's a comfort to know the RIT is there if we need it. One consideration for smaller and rural departments is to train mutual-aid companies in rapid intervention. Cross drills should be held in this concept. These teams can be valuable in those rare but deadly "touch-and-go" fires.

I certainly can justify not assigning a RIT at my normal residential and small commercial structure fires, but I would be hard-pressed to take the witness stand and justify why I failed to assign a RIT at one of those special fires we get every now and then. I knew the risks involved. I was aware of the concept of rapid intervention and the standard that defines it. It's tough to justify my not special-calling for one additional unit when I knew a definite risk was present.

WHO SHOULD BE THE RIT?

Any available unit that has been trained in the concept can serve as the RIT. A good RIT uses some specific tools and actions, but rapid intervention is not a hard concept to grasp. If I didn't have a unit trained in rapid intervention, I would try to get an experienced crew to take this sector. In Toledo, we unofficially trained both of our heavy squads to be the RIT. However, I would not hesitate to call up any company and assign it to rapid intervention if the situation dictated and one of the trained squads was not available.

Figure 18-2. A rapid intervention team represents firefighters' insurance. (Photo by author.)

WHEN SHOULD A RIT BE ESTABLISHED?

Basically, there are two types of fires—those that can be handled using routine measures and those that are nonroutine, where something just does not seem right. Some third-alarm-and-above fires are basically routine in the extinguishment measures required. Some of the most boring fires I have been to were multiple-alarm fires—set up and throw water. On the other hand, an occupancy or structure type could be "nonroutine," perhaps due to the size of the structure involved, the color of the smoke the fire is producing, or some other factor.

Figure 18-3. A fire at which a rapid intervention team is warranted if an offensive attack is going to be made.

Rapid intervention teams should be established at the "nonroutine" incidents or at any incident at which a firefighter could become trapped or cut off by fire. Almost every seasoned firefighter can close his eyes and think of fires at which he was scared, at which he was sure he was going to lose someone, and at which he knew he didn't want to send crews inside even though he believed someone could be in the structure. These are the fires at which a RIT should be assigned. Such fires may include the following scenarios.[3]

- Working fires in a specific area of a single-family or multifamily occupied structure where rescue crews will be required to extend themselves beyond routine measures, such as if there is a report of trapped victims.
- Advanced fires in large commercial structures at which offensive operations are used.
- Shipboard fires.
- Advanced fires in high-rise structures.

All of the above instances have one thing in common: the potential for interior crews to become lost or trapped inside a structure being assaulted by fire. Should interior crews become endangered, the RIT is responsible for locating and removing them.

HOW RAPID INTERVENTION TEAMS WORK

Any unit assigned to be the RIT must understand one thing: It is to act only if a fire company (or individual) is lost or trapped inside the structure. To allow the team to focus on civilian rescue or fire attack removes or divides its focus, which must be specifically directed toward the interior crew's safety. Additionally, the same crew should be used for rapid intervention throughout the incident. It goes back to focus.

The team must be instructed to focus on the following only:

- its mission,
- the tools required,
- its staging location, and
- its on-scene survey.

THE RIT MISSION

The mission of rapid intervention teams is to search for and remove trapped or injured firefighters. That's all. Jim Cline, Ed.D., says it all in "Rapid Intervention Companies: The Firefighter's Life Insurance."[4] Cline points out: Other sectors have specific missions. The same is true of RITs. They are not to be used for victim searches. They are not to be used as exposure or extension sectors. And, although the mission may be similar, they should be separate and distinct from backup. They enter the structure only if needed; and if needed, they get in, get the affected personnel, and get out. This is firefighters' insurance. If I have a claim, I want my insurance agent to focus on my problem and ensure that my needs are met. At a fire, I want the unit(s) assigned to protect me to focus on meeting my needs if I get in trouble.

THE TOOLS REQUIRED[5]

The RIT should carry some specific tools. Apart from the obvious (full turnouts, SCBA, PASS devices, and portable radios), the following should be taken to the command post:

- an extra (full) SCBA bottle for every member of the team;
- door chocks;
- hand lights;
- rescue rope and the necessary tools (carabiners and figure-8 descenders, and so on);
- hauling line or a guide rope, if your department uses either; and
- at least one pry bar, preferably a halligan.

Heavy squad companies could fill a bag with the above equipment, label it "RIT," and store it on the heavy squad. Extra SCBA bottles can be taken from any unit. If the need arises and Command calls for a RIT, the bag then could be grabbed. No time would be wasted looking and checking for the appropriate tools.

WHERE THE RIT STAGES

Once the tools have been collected, the team, if not previously assigned to intervene, should report to the command post or other area designated by Command. If the command post is within easy walking distance of the fire building, it is a logical place for the team to report. Usually the information concerning the structure, the assignment of companies (sectors), and other pertinent issues is available at the command post.

Command (or Operations) may choose to have the RIT stage at a location more advantageous for making quick access into the structure's trouble areas. (At high-rise fires, a more logical staging area for the RIT is in "base"[6] (a

Figure 18-4. An example of the tools required for a rapid intervention team. (Photo by author.)

staging area two floors below the fire). Once at the command post or other destination, the RIT officer should report to Command or Operations that the team is staged and will be conducting an on-scene survey unless it is otherwise directed.

THE ON-SCENE SURVEY

After the RIT has reached its staging area, the final act it is required to accomplish (hopefully) is a quick on-scene survey to determine the following:

- quick ways into (and out of) the structure—fire escapes, exterior stairs, normal interior stairs, pre-positioned ground ladders, and aerial ladders, for example;
- exterior features of the structure—porch roofs, roof type (gable, flat, truss, and so on), ledges, manwalks, window type and configuration; and
- construction type and any positive or negative effect it may have.

While the RIT is staged, the officer and crew can review any preplan drawings or written material concerning the structure and the command board, which should indicate company assignments, sector designations, and locations in the structure. After doing all that, they just wait.

WHAT HAPPENS IF RIT IS ACTIVATED

Hopefully, this "insurance policy" will never have to be used. But if it is needed, some specifics need to be addressed.

- If a crew becomes lost, trapped, or otherwise endangered to the point that it needs assistance in leaving the structure, then Command must first clear the fire frequency of unecessary radio transmissions by asking for "Emergency Traffic."[7] This is not the time to have an engine company calling for 10 more pounds on its 1¾-inch line.

Figure 18-5. A scene survey conducted by a rapid intervention team would note things such as ladders, windows, and fire escapes. (Photo by author.)

- Once Command determines that a crew member or an entire crew is in trouble, he should activate the RIT. Explicit directions concerning the lost or trapped crew—its assignment and location and the number or crew members—must be given to the RIT.
- Command should evacuate all unnecessary sectors from the structure. I can guarantee one thing: If firefighters know that a firefighter is lost, trapped, or injured in the structure—and if they are not "controlled"— they will head en masse toward the area where the firefighter is thought to be. Most times, this is not a good thing to have happen. Freelancing at such a time can become deadly.

If Command or Operations believes that more than one RIT is needed, he can divide the original RIT into groups and assign additional firefighters to work with them. Remember that they have specific knowledge about the lost or trapped crew that other sector officers did not need to concern themselves with until now.

Command or Operations should keep backup companies in the structure. They have a line and personnel who should be relatively "fresh." If the area in which the crew is trapped is being protected by attack lines that are holding the fire in check, then those personnel should be left in as long as possible to help the RIT.

Remember: Unless otherwise directed, only the rapid intervention team searches for and removes trapped or lost firefighters. All other sectors left inside the structure are there to support and protect the RIT. They are not to focus on looking for or removing firefighters.

Rapid intervention is a new concept. It will take time for it to be developed fully. I have presented an overview of the basic concept. If you buy the concept but feel that you need to alter it slightly, so be it. The main thing is that you provide some measure of "insurance" for your members when it is needed. Don't think that this won't be noticed. When your crews know that you have enough interest in their well-being to take out this added "policy" on their lives, they may just go that extra mile for you when it is really needed.

Endnotes

1. NFPA 1500, Section 6-5.

2. Ibid.

3. Although RITs are discussed here only in relation to structure fires, these teams are also used at incidents involving water rescues, haz mats, and confined spaces. The RITs at these incidents are in place before the entry or rescue teams advance.

4. Cline, Jim, Ed.D., "Rapid Intervention Companies: The Firefighter's Life Insurance," *Fire Engineering*, June 1995, 67.

5. Ibid.

6. Many cities that use the incident management system designate the area two floors below the fire in a high-rise as "staging." In Toledo, we found that confusing. Companies already were familiar with staging as it relates to an outside area where uncommitted apapratus/companies gather. With a strong IMS, you should not have unassigned companies anywhere in the building. To us, it just flows better if there is only one staging area. For that reason, we designate the areas two floors below the fire floor as "base." I understand this violates the definition of base and staging, but it works for us and those who respond with us.

7. "Emergency Traffic" is a term that indicates that a firefighter is in trouble and that all unnecessary radio traffic must cease. Command must call for and terminate "Emergency Traffic." This term came to my attention in *Fire Command* by Alan V. Brunacini (NFPA Publications, 1985). In Toledo, we use "Emergency Traffic" just as they do in Phoenix.

REVIEW QUESTIONS

1. What is the mission of the rapid intervention team?

2. Command may consider establishing a RIT if he suspects what two things?

3. What standard calls for the use of a rapid intervention team?

4. What are the qualifications for rapid intervention team members?

5. At which four types of incidents should Command consider the use of a RIT?

6. The RIT must understand that it is to act only under which conditions?

7. List the tools required for rapid intervention.

8. What are the components of the RIT's scene survey?

9. Where should the RIT stage?

10. What would be the function of the working crews inside a structure if rapid intervention were activated?

DISCUSSION QUESTIONS

1. Compare and contrast rapid intervention with backup sectors.

2. Discuss the concept of "Emergency Traffic." Can sector officers be expected to follow this rule if activated? What options does Command have for violations of this concept in your department? Is there or can there be any discipline tied to a violation of Emergency Traffic? What would happen if an officer were to give a nonessential transmission while you were under Emergency Traffic?

3. Refer to question 10 above. What exceptions would apply in this situation? Why would they be made?

19

Benchmarks

When we implemented the incident management system in Toledo, we, like many other cities, experienced some adjustment pains. Some of the things that we were trying to accomplish seemed so simple, and yet we were still grappling with them. For example, we kept getting complaints that too many people were inside a single-family home at the typical fire and that backup crews were looking for and putting out fire.[1] We were making all of the appropriate assignments (the Big Four: attack, ventilation, search, and backup), but found that there was still a lot of confusion at the fire scene. It seemed that perhaps the system that was meant to be the prototype for incident management was flawed.

After discussing this issue with some line battalion chiefs, we believed we had stumbled across the problem. As simple as it sounds, here was the problem: We were sending crews inside a structure now with

KEY CONCEPTS

✔ "All clear"
✔ "Backup line in place"
✔ Benchmarks
✔ "Crew located"
✔ "Crew outside"
✔ "Exposures covered"
✔ "Extension areas checked"
✔ "The fire is knocked down"
✔ "Overhaul complete"
✔ "Salvage complete"
✔ "Under control"
✔ "Ventilation complete"

specific objectives (i.e., sector assignments). They had an obtainable goal and were pleased to go in and work toward the accomplishment of that goal. The problem was that, in the past, prior to the incident management system, most officers and their crews went into a building based on (1) the type of apparatus they were riding; (2) any information given them by a superior officer (for lack of a better term, an assignment); (3) any specific SOP[2] that our department had developed; and (4) whatever it was that they felt needed to be done. When that was completed, they did the next thing they considered to be necessary. Now, all of a sudden, we sent them into a building and told them (and I'm paraphrasing): "Here's what I want you to do. Go do it, and then don't do anything else until I tell you. "When they were done with their assignments, they weren't sure what we wanted them to do next.

One more problem compounded the situation. Chief officers were "chained" to the command post. Not literally, of course; but we were dissuaded from leaving the outside to go inside and "take a look around." The outcome was that inside the building were too many people who had completed their assignments and were waiting until someone told them to come out or until the SCBA bell went off. It may sound silly, but that's what was happening. We found the solution to this problem—benchmarks.[3]

BENCHMARKS

Benchmarks are announcements that a particular activity or assignment has been completed. They serve three purposes. First, they let Command know that a specific activity has been completed. For Command (or Operations), who is outside, this information enables him to eliminate items that are on the to-do list and move other items up on the list. Command may have no other way of knowing when some of these particular activities are complete.

Second, benchmarks lend an air of "closure," if you will, to a sector assignment. Time announcements or notations should be made to indicate when specific activities have been completed. If legal actions or questions concerning the incident arise, the incident time line can give a relatively true picture of the incident.

Third, benchmarks are designed to give Command a better understanding of the progress being made. Again, Command can't (and probably shouldn't) go inside. If Attack gives the benchmark indicating that the fire has been knocked down and Command still sees open flame in the area of the attack sector, he will be aware that Attack may not have gotten all the fire and that there's something Attack's not seeing. If Ventilation indicates that the ventilation hole is opened up and the interior attack company still can't make the second-floor stairs, that indicates something is stopping the heat from going up and out the vent hole and that that condition will have to be corrected.

Benchmarks provide a systematic "check-and-balance" system that permits Command to determine whether what sector officers believe to be happening is indeed happening. Benchmarks are brief and specific. Those used in Toledo are given below. You can use some, all, or none of them; you can make up your own. Just remember the following:

- They should be brief. There's no need to tie up radio airtime with a lot of words.
- They should be uniform. The benchmark for ventilation should always be the same, no matter which form of ventilation is used. This makes it easier for the officer and Command to understand exactly what is being said and what it actually means.
- They should be used all the time and by virtually all sector assignments.

Command's Benchmark

"Under control" is the benchmark given by Command to Dispatch when conditions warrant. It basically is the only benchmark Command need give. It indicates the following:

- The fire is under control or the major portion of the incident is over.
- The need for additional equipment or mutual aid no longer exists or has been substantially diminished. Some departments put other units (or other departments) on standby during incidents. This benchmark could serve to let the standby units stand down.

Attack's Benchmark

When Attack gives the benchmark "The fire is knocked down," it indicates that Attack has found and knocked down the main body of fire. Attack gives this benchmark to Command as soon as practicable. As indicated above, if Attack gives his benchmark and Command still sees fire showing or heavy volumes of black smoke pushing out from the structure, he knows something is wrong.

After Attack gives this benchmark, Command can respond "Okay" and say nothing else. In that case, Attack may begin to check for extension if he believes that action is required to complete his mission. Once

Figure 19-1. Command is the individual on the scene who gives the "Under control" benchmark. (Photo by author.)

the structure has been checked for extension (if necessary), Attack should begin to overhaul the fire area. However, note that if no additional assignments are made that connect attack to extension or overhaul, the time line is disrupted and, as far as an attorney is concerned, extension and overhaul have not been done. Get used to making the new assignments.

Remember that Extension has only the specific task of checking for additional fire. His crew should not put out the fire unless Command directs them to do so. Overhaul is responsible for putting out the last vestiges of the fire, no matter where it is, and for determining the area of origin. The normal progression is for Attack to go from Attack to Extension to Overhaul or to have Command reassign Attack to be Overhaul (or Extension)[4]:

Command to Attack: I'm changing your designation to Overhaul.

Now Command no longer has an attack sector. He has a crew whose focus is to put out the remaining fire. It must be kept in mind that the benchmark given by Attack ("The fire is knocked down") does not mean that the fire is out—just knocked down.

A final note: As Command, I hesitate to give an "Under control" immediately after I get a benchmark from Attack. A lot of things may still be going on, and it's embarrassing to have to start special-calling for equipment after you have given an "Under control."

Search's Benchmark

Search's "All clear" benchmark, given to Command immediately on completion of the primary search, indicates that a primary search has been conducted and that all savable victims have been removed from the structure. (Command should repeat this benchmark to Dispatch for the incident time line.)

Several specifics must be noted in relation to searches, benchmarks, and incident time lines. First, you can have several DOAs at an incident and give an "All clear." By design, this benchmark indicates that all savable victims have been removed from the structure. If on arrival the structure is in total flashover on all floors, a search should still be done. It will not be on the top of Command's to-do list because it should be evident that no savable victims are in the structure. If on arrival the structure is heavily charged with thick black smoke and responders have been on the scene for 10 minutes, there is little need for a primary search. For both of the above scenarios, a secondary search is indicated later in the operation, when it can be done safely and effectively. If on arrival it is not known whether anyone is in the structure and fire conditions indicate that savable victims may be inside, a search sector is warranted. Once the search has been completed and the "All clear" has been given, Command can assign Search to start a secondary search, reassign Search to another sector, or bring the crew out of the structure.[5]

Remember the complaint of having too many people in the building, as discussed at the beginning of this chapter? This approach helps to control that situation.

"Everyone Is Out" vs. "All Clear"

A problem we ran into when we first used benchmarks in Toledo was that civilians or police officers would come up to the first unit on the scene and report, "Everyone is out of the building." Some of our officers were confusing this with an "All clear." That's wrong. An "All clear" means specifically that a search sector was assigned and that they had entered the struc-ture and had completed a primary search. An occu-

Figure 19-2. *Search will give an "All clear" after conducting a primary search of the structure.*

pant or the police's saying that everyone is out of the structure means noth-ing—absolutely nothing—even if the occupant is the one telling us everyone is out and that he is the only one who lives there. What if a neighbor should run in the back door looking for the occupant and no one knows this? What if a police officer had attempted a rescue prior to our arrival? What if the fire had been deliberately set and the arsonist was still in there? If someone is out front telling you that everyone is reported to be out, then state just that!

> *Command to all units and Dispatch: The police report that everyone is out of the structure.*

Now Command can make some assignments, keeping in mind that a search sector still needs to be on the to-do list, although possibly not at the top.

Backup's Benchmark

The benchmark "Backup line in place" is given by Backup to indicate that the backup line has been pulled, stretched, and charged in the appropriate area in the structure. Backup gives this benchmark to Command immedi-ately on placement of the backup line. Once Command knows a protective

Figure 19-3. If civilians report that all the occupants are out, a primary search still must be conducted. The position of search on your to-do list may change.

line is inside the structure for the sole purpose of keeping an eye on the working crews, he can think about the possibility of "extending" things a little more. It is imperative that Command be informed when the backup line has been positioned, not when it is being positioned. It all goes back to focus. If the officer assigned to backup is in the process of looking for the appropriate location for the backup line, he hasn't had adequate time to focus on much else. When the backup sector gives the benchmark, Command knows that the backup sector is ready to focus on the safety of interior crews.

Ventilation's Benchmark

When Ventilation gives Command the benchmark "Ventilation complete," it indicates that natural or mechanical ventilation has commenced or that an adequate ventilation hole has been opened on the roof or in another appropriate area.

Note that this benchmark is not given when all the smoke and other by-products of combustion have been exhausted from the structure or area. It can (and should) be given as soon as the mechanism of ventilation is in operation. I have been at fires in large industrial structures where the command post was so remote from the area in which the positive-pressure fans were operating that I could not hear them running. All of a sudden, I observed smoke doing things that it hadn't been doing two seconds before. This caused concern. Most of the topside ventilation is done away from Command's

Figure 19-4. Once Ventilation opens the roof, he should give the benchmark "Ventilation complete" to Command. Smoke should be coming from the hole after the benchmark has been given.

view. Command knows that he ordered ventilation. Command knows it is being accomplished by topside ventilation (there are ladders going up to the roof). If the benchmark is given by Ventilation, Command knows that the process has commenced and conditions should be changing; otherwise, something is wrong.

Exposure's Benchmark

Exposure gives Command the benchmark "Exposures covered" as soon as protective lines have been placed and are in operation.

Note: Exposure will then, if necessary, enter the exposed building to determine whether fire has entered the structure. This fact should be relayed to Command as soon as entry has been made.

Once Command hears an officer is looking out for the exposed structures, he can again "sit back and look for other problems to solve."

Extension's Benchmark

Extension relays to Command the "Extension areas checked" benchmark as soon as extension has checked the area surrounding the fire. Command now knows that an officer and his crew checked above, around, and below the fire for extension. Had the extension officer found any extension of fire, he would have informed Command of that fact, as well as of the location of the extension, and let Command (or Operations) determine how best to control it. Normally, Attack handles the extension of fire and extinguishes it while Command either reassigns Extension or tells him and his crew to get out of the structure.

Overhaul's Benchmark

The benchmark "Overhaul complete" is given to Command by Overhaul as soon as the area of origin has been determined and the last vestiges of the fire have been extinguished. It should be given prior to removing the last line within the structure and after the overhaul officer has taken his last walk through the structure to look for any traces of smoke or fire.

Salvage's Benchmark

The "Salvage complete" benchmark is given by Salvage to Command after all savable property has been protected from the effects of the fire. This benchmark does not mean that the ceiling has stopped dripping and the tarps can be removed. It means that the tarps have been spread and are keeping water off valuables.

Rapid Intervention Team Benchmarks

The benchmarks "Crew located" and "Crew outside" are given by the rapid intervention team to Command at the appropriate time. "Crew located" indicates that the lost or trapped crew members have been found. It says nothing about their condition. The benchmark "Crew outside" is given when the crew has been taken outside. Remember that RIT might choose to remove the crew by a route that may be out of view from the command post. God willing, this will be the extent of the benchmarks and similar announcements that the RIT has to make.

WHAT BENCHMARKS MEAN

Benchmarks mean the end of incident confusion. Areas that needed to be visually checked by Command in the past now can be verbally "checked" by systematic, standardized statements.

Once Command is informed that an activity has been completed, he can reevaluate the scene and his to-do list and do one of the following:
- give that crew a new assignment;
- have that crew report to another sector under another officer; or
- have that company report outside for a break, go to rehab, or go back home.

It's Command's choice. Benchmarks help give Command total control of the incident.

Endnotes

1. That problem was solved by the use of mission statements.

2. Some or most departments at this time had specific standard operating procedures (SOPs) that were followed at most incidents. Some SOPs were very vague, whereas other cities had specific, definitive procedures that gave a litany of "things" a unit had to do under certain circumstances. Although some departments still rely heavily on SOPs, I believe the use of the incident management system decreases the need for such structured procedure lists.

3. Brunacini, Alan V. *Fire Command*. 1985. National Fire Protection Association Publications, 70.

4. Review the mission statements of Attack, Extension, and Overhaul if you're confused about the responsibilities of each.

5. Several departments pull the search company out of the structure after search has been completed and use it as a rapid intervention team.

REVIEW QUESTIONS

1. Define benchmarks.

2. What three purposes do benchmarks serve?

3. What characteristics should benchmarks have?

4. What benchmarks does Command normally relay to Dispatch?

5. Where should all benchmarks be noted?

6. Match the appropriate benchmark with the sector:

   a) "Under control"   ___ Rapid intervention team
   b) "The fire is knocked down"   ___ Salvage
   c) "All clear"   ___ Exposure
   d) "Backup line in place"   ___ Attack
   e) "Ventilation complete"   ___ Search
   f) "Exposures covered"   ___ Overhaul
   g) "Extension areas checked"   ___ Ventilation
   h) "Overhaul complete"   ___ Backup
   i) "Salvage complete"   ___ Extension
   j) "Crew located"   ___ Command

DISCUSSION QUESTIONS

1. What are some alternate benchmarks for indicating that a sector's task has been completed or is in operation? Compare your list with the author's.

2. Do benchmarks have relevance for Command's to-do list? If so, explain how.

3. Discuss "All clear" in relation to the finding of fire victims. Should an "All clear" be given if you do have fatalities, or should another term be established?

Section IV

Miscellaneous Incidents

Emergency Medical Incidents

Routine emergency medical service (EMS) runs under informal and formal command (you may want to review Chapters 2 and 3) are discussed below. Mass-casualty incidents are not included.

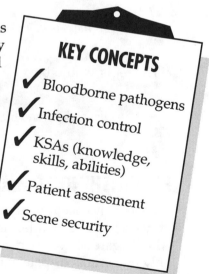

KEY CONCEPTS

✔ Bloodborne pathogens

✔ Infection control

✔ KSAs (knowledge, skills, abilities)

✔ Patient assessment

✔ Scene security

ROUTINE FIRST-RESPONDER INCIDENTS—INFORMAL COMMAND

Fire departments that respond to EMS incidents dispatch engine companies, rescue squads, or rescue units (smaller two-person units). In Toledo, we send a truck company on an ALS (advanced life support) run if it is the closest unit available.

ON-SCENE REPORT

When dispatched on a typical EMS incident, a first responder reports that he is on the scene in the same manner as he would for a still-box fire incident. This on-scene report accomplishes two functions: It adds information to the incident

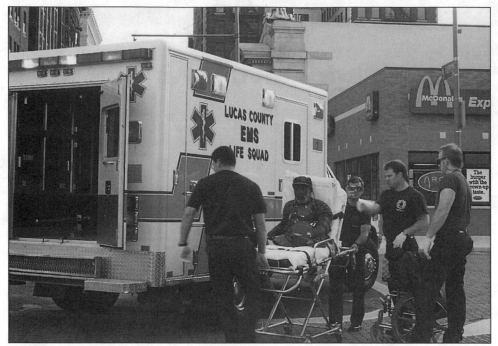

Figure 20-1. Crews operating at an emergency medical incident. (Photo by author.)

time line and provides a method for checking that the crew is reporting to the correct address. The on-scene report would be similar to the following:

> *Engine 3 is at 132 River Place.*

The reply would be

> *Dispatch: Okay, Engine 3, 1323 hours.*

ESTABLISHING COMMAND

There is no need to announce the assumption of command when there is only one first-responder unit, whether it is a BLS (basic life support) or an ALS. Regardless of the type of incident, command need not be established if only one unit is dispatched.

Informal Command

When one officer responds, informal command is used. The dispatching of paramedic units should not change this fact. In some departments, the paramedics are not employed by the fire department or any city agency. In other departments, the paramedics are fire personnel with additional training and have been certified as medics, but paramedics usually do not hold rank. The fire officer responding is Command (informal). As such, he is in charge of the scene. If paramedics are also dispatched, they (in some states by law) are in charge of the victim(s). Some lines may be crossed here. There may be some questions concerning the well-being and safety of the victim. I do not intend to get into these scenarios. If there are questions or problems concerning the lines of responsibility between fire officers and paramedics (or ambulance units) in EMS responses, your fire administration or legal department will need to address these concerns. Procedures must be established so that all units or entities responding understand where their responsibilities and authority lie.

Victim Care

As with any incident, Command's responsibility—whether under informal or formal command—is to focus on the whole incident. Typical EMS incidents are no different. The officer should avoid being a hands-on guy if at all possible. I realize that at some incidents the officer must participate in patient care for the welfare of the victim. In those instances when the officer must work with the victim for an extended period of time, it is vital that he get additional help so that he can again pull back and focus on the incident.

As at a fire incident, an officer at an EMS incident should not make assignments over the radio when informal command is used. That would be a

Figure 20-2. Informal command at an EMS incident—one unit, informal command. (Photo by author.)

waste of airtime. At these instances, a set of assignments must be made. The assignments—"Bill, on EMS runs, you take vitals; Mary, you set up the oxygen," for example—could be made at roll call to avoid confusion. (Protocols from medical directors or other trained certifying entities will dictate the measures required for specific medical problems. I do not have any special KSAs (knowledge, skills, and abilities) that allow me to dictate medical procedure. I feel confident, though, that the incident management system fills the void as it pertains to roles and responsibilties at EMS incidents.

The following four items should be on the to-do list for a typical EMS incident:

- *Meet the crew's needs.* Things are not as they once were. When I was a line officer, things were somewhat easier. Bloodborne pathogens, infection control, rubber gloves, and assaults on firefighters were unheard of. We never "gloved up." We did mouth-to-mouth rescue breathing—it was part of the job! (The recruit got the mouth.) We rarely feared for our own safety. No one disliked the department. Basically, we were off-limits to the "bad guys." As is the case at a fire, one of Command's priorities at all EMS incidents is firefighter safety. If the officer in charge notices that members are not "gloved up," he should take steps to protect them.

  All areas of the country are seeing an increase in gang violence and drug-related crimes. It is better if the officer is free to stand aside and observe the whole scene while his crew members work on the victim. It can all happen so quickly. Someone should just stand back and watch our members' backs.

- *Meet the victim's needs.* Taking vital signs, supporting with oxygen, assessing patients, and providing emergency treatment fall under this category. Normally, the officer will have crew members handle this aspect of the incident.

- *Transport of the patient(s).* Not all EMS victims need transporting. I spent a year as lieutenant on the busiest rescue squad in the city and made my share of "goof" runs. Some departments provide ambulance service, and this may not even require a "step"—(just load and go). However, only 30.55[1] percent of departments that provide EMS provide their own means of transporting. The remainder of the departments that provide EMS use private ambulances or some other means of transport. After the victim has been assessed, the officer or paramedics should determine the level of transport (ALS, if available, or BLS) needed and call for the proper unit. Regardless of the type of system under which a department operates, legally it is in our best interest to ensure that victims are referred to an appropriate health-care facility.

- *Secure the scene.* Once the patient is stable and en route to a health-care facility, the officer has the responsibility to secure the scene. This task has many aspects. If a crime was involved—such as an assault (with or without a weapon), robbery, domestic violence, and so on—and the police or other law enforcement agency that has jurisdiction has not been notified, the officer

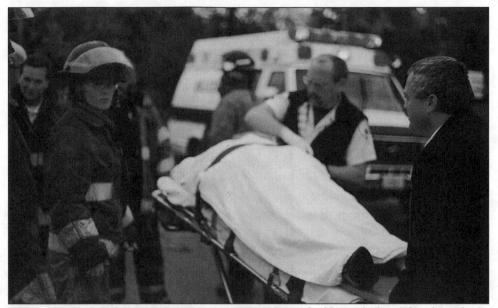

Figure 20-3. An ambulance preparing to transport a victim to a health-care facility. (Photo by author.)

must make the appropriate notification and, in some instances, preserve the scene by standing by until the police arrive. Weapons must be left where they are found and not be distrubed. If the victim is DOA, the scene probably will have to be turned over to the police.

Another aspect of scene security is notifying the victim's family. The security of the residence (if the incident occurs at home) is also the responsibility of the officer. An officer would find it difficult to defend leaving an apartment or home open when leaving the scene after the victim has been transported. Normally, if the victim is conscious, he will tell you how to lock up and where to put the keys. Most times, a neighbor can be trusted to secure the residence. (Check with the victim before entrusting a neighbor. Also, note on the run report and in a company journal who was left in charge of the residence.)

Finally, it may be advisable to take a quick walk through the house with a crew member to make sure that the stove is off and that any other sources of ignition are removed. The victim, if conscious, can help reduce the time spent checking sources of ignition by responding to a few quick questions.

I have been on enough EMS runs to realize that additional tasks may be needed at some EMS incidents, but I believe that these four items constitute the beginning of a generally good to-do list.

Formal Command

Occasionally you will be dispatched to more complex EMS incidents (other than vehicular accidents) that will require additional personnel from the onset. These incidents in-clude shootings, bar fights in progress, persons trapped or impaled by machinery, and the like. In these in-stances, more than one unit is dis-patched (with officers in each unit), and someone must take the role of Command.

The On-Scene Report

Once dispatched, the first unit to arrive on the scene establishes Command as part of an on-scene report, which should contain three parts that will serve to

- establish the on-scene time that will be reflected in the incident time line,
- verify the correct address, and
- formally identify Com-mand.

The transmission would be something like the fol-lowing:

Figure 20-4. An incident at which formal command is used at an EMS incident. (Photo by author.)

Engine 6 at 243 Main Street. Engine 6 is Main Street Command.

Okay. Engine 6 on-scene, and Main Street Command at 0344 hours.

From this transmission, we now have the following:
- Engine 6 is on the scene, and another "mark" can be noted on the inci-dent time line.
- Engine 6 is at the location to which it was dispatched. (It may not actu-ally be the "correct" address, but the engine is where it was sent.)
- The Engine 6 officer is Command.

Now, the same rule applies at a fire incident. Once someone for-mally establishes command, all other units responding should stage. This rule needs to be maintained at all types of incidents. Usually the second first-responder will not be needed anyway. At about the time the second unit comes tramping inside, Command gets on the radio and cancels it. If it is not canceled, just about the time the crew members arrive upstairs (or inside), Command will be asking them to bring in the backboard and cot. No matter what the reason may be, if Command is to have total control at a two-unit-or-greater response at EMS incidents, all other responding units should stage and await direction from Command once formal command has been established.

A final thought concerning EMS incidents: Other than for vehicular accidents, I am not aware of any "formal" sector assignments for basic EMS incidents. At mass-casualty accidents and large vehicular accidents, specific sector assignments such as triage, treatment, transport, extrication, and so on can and should be made. When there is more than one victim, I believe that specific patient care should be assigned. Consider numbering or tagging victims.

 Command to Engine 13: Take the stabbed man in the blue shirt. Police will show you where he is.

Any more than that may be unnecessary as well as confusing. Once Command has made patient assignments, he can properly manage the incident and even reassess victims to ensure that no one or nothing has been missed.

Endnote

1. 1993 Fire Department Survey, Phoenix (AZ) Fire Department.

REVIEW QUESTIONS

1. At a normal EMS run with first responders and paramedics, what is the responsibility of the officer on the first-responder unit? What is the responsibility of the paramedics?

2. List the four items that should be on the to-do list for EMS incidents.

3. What is one of the concerns about meeting the needs of the crew?

4. You are the officer of an engine company responding to a shooting in the west end of the city. Also responding is another engine and a medic unit. On arrival, should formal command be established? ___ Yes ___ No

5. In the above situation, what would you expect the second unit to do when it arrives at the scene?

DISCUSSION QUESTIONS

1. Discuss the responsibilities of the officer on a first-responder unit and the paramedics at an EMS incident. Who should have the final say in any dispute?

2. Discuss the concepts of scene security and securing the structure. In your opinion, is this the responsibility of the fire department?

3. In your opinion, is staging at an EMS incident necessary or just a waste of time?

21

Vehicular Accidents

TOPICS

- **The differences between a fire incident and a vehicular accident on-scene report**
- **Staging at vehicular accidents**
- **Staging for transport units**
- **Using sectors at vehicular accidents**
- **Specific mission statements of sectors assigned at vehicular accidents**
- **Types of extrication**
- **Responsibilities of sector officers at vehicular accidents**
- **Benchmarks used at vehicular accidents**

The incident management system (IMS) is well suited to vehicular accidents. Control can now be the rule where once chaos often reigned. Prior to the IMS, no clear-cut control was established at many vehicular accidents. At times, victim and personnel safety were neglected and requests for transport vehicles were conflicting.

Under the IMS, the same principles that hold for structure fires hold for vehicular accidents: Freelancing is not to be tolerated, and the first unit on the scene must establish command. From that point, Command assigns the appropriate sectors. (A vehicular accident should be sectored in much the same way as a structure fire.)

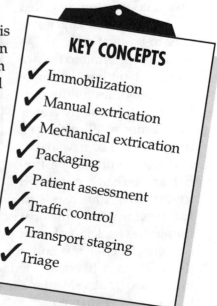

KEY CONCEPTS

✔ Immobilization
✔ Manual extrication
✔ Mechanical extrication
✔ Packaging
✔ Patient assessment
✔ Traffic control
✔ Transport staging
✔ Triage

ON-SCENE REPORT AT A TYPICAL VEHICULAR INCIDENT

Several years ago, our department struggled with the response for vehicular accidents, specifically with regard to victim and firefighter safety. I was part of a committee that developed the new dispatch procedure. The problem was the training we were receiving from our EMS Bureau. They maintained that victim care for vehicular accidents needed upgrading. All victims transported needed to be "boarded," and all boarded victims needed collars.

Next came the towel rolls— two standard bath towels, rolled and taped into approximately 14- × 18-inch rolls, are placed one on each side of the victim's head so that the head is immobilized on the backboard. Commercial head/neck immobilization devices are available, but homemade towel rolls work well for us. Patient care for victims involved in vehicular accidents was becoming very labor-intensive. We developed a rule of thumb that established as a *minimum* two EMTs per noncritical victim[1] and two EMTs and one paramedic[2] for every critical victim. The op-timum called for a somewhat larger rescuer-to-pa-tient ratio.

We also determined that we were not pulling adequate protection lines for fire control at accidents. My fear is that someday I'll pick up a trade journal or newspaper and see the photo of a firefighter who had been burned

Figure 21-1. A firefighter providing C-spine immobilization during packaging. (Photo by author.)

to death with his hands in position to provide C-spine immobiliation. He had been frozen by the heat of a flash fire that trapped him while he was sitting in the backseat of a car providing C-spine immobilization for a victim. No one pulled a line, and, for some reason, a flash fire occurred. Hopefully, this will never happen. The fact is, however, that we do not pull protective lines enough at vehicular accidents. If they do get pulled, too many times they don't get charged.

Figure 21-2. What information would you include in an on-scene report at this incident? (Photo by author.)

We send a minimum of two units (five firefighters and two officers) to every vehicular accident. We also ensure that one of these units is a full-size engine that will provide backup lines. Using the rules of thumb for victim care and having one officer take command, this response enables us to provide basic life support (BLS) for two victims.[3]

The first unit arriving at the scene of the accident should give an on-scene report that provides much the same information that would be given at a structure fire. The report for a vehicular accident should contain the following components:

- *The fact that the unit has arrived on the scene.* As previously noted, this information is entered into the incident time line.
- *The fact that the unit is at the proper location.* Many times, crews reported to the wrong intersection or were dispatched to the wrong location as a result of Dispatch, the 911 operator, or caller error.
- *A brief report of observable conditions.* It is understood that this size-up may not be accurate, but key phrases or vehicle counts will help Dispatch, as well as other units responding and listening, to get a handle on the probable magnitude of the situation. Included in this conditions report should be the following:
  —The number of vehicles involved. This information gives an indication of how many units may be needed. As a rule, we tell our officers to call initially for one unit per vehicle involved. This goes back to the rescuer-victim ratio. If, after a quick triage of victims, it is determined that the injuries are minor, called-for units can be canceled.
  —If vehicles are not upright.
  —An approximate count of the number of victims, if possible. (It may

not always be possible to do this, especially if it is dark or there is heavy fog or some other vision-restricting weather condition.) An approximate count of the victims will signal dispatchers to consider whether they need to have additional help available should Command ask for it.

- The establishment of some type of command according to the number of units initially dispatched or needed. (If the first unit to arrive finds only one vehicle is involved, the vehicle has only minor damage, there is only one passenger, and injuries do not appear to be serious, the officer may cancel any other responding units and enter into informal command. Another possibility is that the first unit on the scene may need to establish formal command.

Initial On-Scene Reports

Examples of initial on-scene reports for two-engine responses follow:

Dispatch: Engine 17 is at Central and Collingwood. We have two cars involved. Looks like three victims. Engine 17 will be Central Street Command.

Dispatch: Engine 13 is at Front and Consaul. We have a one-car minor accident. Cancel Engine 6.

Once Engine 6 has been canceled, the only unit left at the incident will be Engine 13. That brings us to an informal-command situation, which requires no specific announcement of the assumption of command.

Dispatch: Engine 5 is at Washington and Erie. We have a four-car accident with some heavy damage. Engine 5 will be Command. Send me two more engine companies and a medic unit.

STAGING

As at structure fires, once someone formally establishes command, all other units responding should stage. Except for the extremely large accident (mass casualty) that requires multiple alarms, units responding to vehicular accidents will work under the constraints of Level I staging (this does not apply to transport units, which will be discussed later). Units responding will stage at a location that does the following:

- Provides the best access to the accident location. Consider that the location of the vehicles involved, the police cars (which may be parked anywhere), and already pulled backup lines may force responding units to go around the block for easier access to the accident.

• Allows access for other responding *special* units such as transport units and extrication squads. If you know they are coming and you know you will be in the way, hang back to allow incoming *assigned* units access to the area.

Staging for Transport Units

One of the problems associated with vehicular accidents before we began using the IMS was that responding transport vehicles would come right up to the scene and get the cots out. About the time the victim was being pulled from the car, the ambulance attendant would be shoving a cot in the backs of our knees. Staging should eliminate scenes such as this.

When Command realizes that he will need more than one transport unit—whether a medic unit with ALS (advanced life support) capabilities or a BLS private or public ambulance—he should establish a Level II staging area for the incoming transport vehicles, especially if extrication is in progress and transport will be delayed.

Command should inform whoever is dispatching transport units to have all responding transport units stage at a specific location. A firefighter can be sent to that location with a portable radio to take on the duties of staging officer. When an officer assigned patient care informs Command or Operations that a patient is packaged and ready for transport, a transport vehicle can be called up and directed to the patient's location.

 Command to Staging: Send me one ambulance. Have it pull behind the green station wagon in the middle of the intersection.

Transport staging works in the same manner as staging for a fire. It provides control of the accident scene: Command controls when and where ambulances and other transport vehicles report. Police officers must be informed of this procedure. As Command, I have seen them wave up ambulances responding to staging before I asked for them. They need to know what staging is and how it works.

INITIAL TRIAGE

At a vehicular accident that involves more than one victim, the first task is to triage victims quickly. I once believed that this should have been done by Command as part of his size-up. Now I believe that another member of his crew should be assigned this task. The first officer on the scene (Command) must be able to stand back and focus on the whole. Moving quickly from vehicle to vehicle or around a single car to triage victims visually is not beneficial to the incident as a whole. As is the case in any other incident, Command must focus on the entire scene to ensure that nothing escapes him. Aspects such as downed wires, other involved vehicles, victims ejected from vehicles, and crowd control may be missed and can result in fatalities. At mass-casualty incidents, triage is a formal sector assignment. Once on-scene, a responder with a higher level of training in emergency medical response should reassess the triaged victims and adjust patient care as necessary.

SECTORS AT VEHICULAR ACCIDENTS

The final piece of the puzzle in controlling and effectively managing vehicular accidents is sectorization (presented in Chapter 8). The basic types of sectors assigned at vehicular accidents—and all other incidents—are as follows:
- functional—defines specific tasks to be accomplished,
- geographical—defines specific areas in which to work, and
- combination—involves functional and geographic sectors.

Functional Sector Assignments

Following are the sectors most commonly assigned at vehicular accidents, plus their mission statements.

The Mission of Patient Care

Patient Care's mission is to provide for the emergency medical needs of victims in the area assigned. Patient care, probably the most commonly assigned sector at a vehicular accident, basically encompasses four areas.
- *Patient assessment* (the ABCs—airway, breathing, and circulation). If more than one victim is within the assigned area of responsibility, this

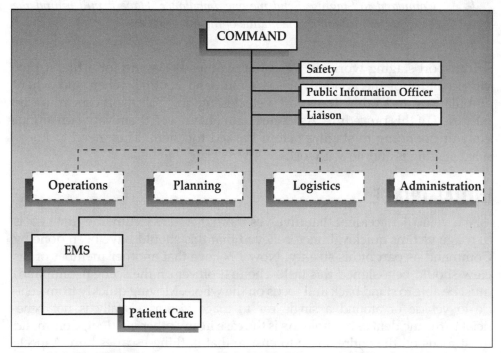

Figure 21-3. The relationship between Patient Care and other sectors at a vehicular accident.

assessment should also include basic triage to determine which victim has the most urgent need. Also, primary and secondary assessments must be made and vital signs taken and retaken at specific intervals.

- *Immobilization/treatment.* Cervical collars should be mandatory for any victim who is to be transported. Splints or other immobilization devices should be applied as needed. Bandaging should be part of this step if there are open wounds. If the victim is to be supported by oxygen, it should be started at this time. (If ALS tactics are indicated, they are performed during this phase of patient treatment.)

- *Packaging.* If a victim is to be transported, he must be properly packaged. Collars are mandatory. If a collar is indicated, then a backboard is also indicated, as are towel rolls or other forms of head/neck immobilization. Victims are secured to the board and placed on a cot in preparation for transportation.

- *Transport.* The victim must be transported to a health-care facility for treatment.

Most departments that provide EMS response within their jurisdiction handle the assessment, immobilization, and packaging. The transporting of victims may be done by the department; a private service; or a county, city, or other agency, depending on the organization of the department and any contractual arrangements it may have.

When patient care is assigned at an accident, it must be announced over the radio (as a sector would be announced at a fire). Doing this lets everyone know that there are (or may be) injuries at this incident. Remember, we are discussing incidents to which at least two units have been dispatched. At smaller incidents, only one unit may be sent. Dispatch may have reason to send more than one unit—even if there are no victims—based on the information received from the caller or 911 operators. Examples would be incidents involving the spill of a hazardous material or the downing of an electrical wire. Higher-ranking officers, specifically chief officers, now know that there are probable injuries at this incident. (They hear that Command is focusing on the incident.) The announcement also enables the unit being assigned to determine the tools its crew should carry to the scene—EMS equipment, backboards, a hoseline for backup, or lunch? Command's assignments must be specific if the incident is to flow as smoothly as possible. *Remember:* We were sent to bring control and order out of chaos, not to provide additional confusion.

As stated earlier, the initial rule of thumb is to provide one company (unit) per vehicle involved. If more than one vehicle is involved, the assignment should be a specific unit to a specific vehicle. If the personnel in any company can handle the needs of their victims with personnel to spare, then Command should be notified so that personnel can be assigned elsewhere. If several victims are in the car and more help is needed, the officer assigned to that sector must inform Command so that additional help can be provided.

Engine 4 is at Hill and Westwood. Two cars are involved in a minor accident. Engine 4 will be Hill Command. Engine 4 will be Patient Care in the brown Ford. Engine 14, you're Patient Care in the green Pinto.

That may seem like a lot to say at a small accident, but the transmission took only 13 seconds, and it defined the roles and responsibilities. If other sector assignments were needed, they should have been included so that everyone would know what everyone else is doing.

Once all the victims in an area of assignment have been assessed, immobilized, packaged, and transported, the sector officer should inform Command by giving the proper benchmark ("Victims treated"). Command will reassign the sector or have the members start to pick up and go back. It's Command's decision.

The Mission of Backup

Backup's mission is to provide protection to personnel and victims in case a flash fire should occur. He does this by pulling, strategically placing, and charging a hoseline of adequate diameter. If this is done at every accident at which vehicular fluids are spilled or mechanical extrication is needed, we may never have to read about the scenario I described earlier, in which a firefighter burned to death while in the backseat of a vehicle while providing C-spine immobilization. A line capable of providing the required gpm must be placed where it will protect personnel and victims.

Some specific rules apply to the backup sector:
- The backup line should be 1½ inches. Flows of at least 95 gpm may be

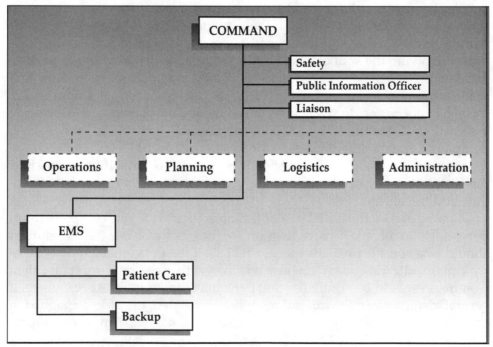

Figure 21-4. The relationship between Backup and other sectors at a vehicular accident.

required to push a flash fire of gasoline off personnel or victims.

- Any line pulled must be charged. It is unacceptable to pull a line and not charge it. The time lost could equate to lives lost.
- This same rule applies to handling the backup line. Leave a man in full turnouts holding the charged line.
- Everyone needs a focus area. Backup's focus is on being prepared with a charged handline in case of flash fire. Command oversees the entire incident. Patient Care attends to the needs of the victim. None of these areas can receive the attention they must have if the officers assigned to them have their heads inside of car windows checking pulses or are calling for another ambulance to come up from staging.
- Backup's line should be positioned so that he would be able to cut off the spread of any flash fire and to protect anyone nearby. He should be close enough to be effective but not so close that he is in the way or may become involved in the fire himself. Backup must make sure that he isn't standing in spilled gasoline.
- If the fire potential is great and personnel are available, a firefighter should be left at the pumps.

Figure 21-5. A backup line pulled, charged, and held at a vehicular accident. (Photo by Ken Anello.)

Calculated risks are part of our job. Once the line has been pulled and charged, it should be tested to make sure that water comes out of the tip and that the nozzle is on the proper setting (30° fog or slightly larger). Whether the pumper driver should take on other duties is a judgment call to be made by Command and the officer assigned to backup.

After the line has been pulled and stretched and the driver question has been resolved, the remainder of the crew can be assigned to another sector. The officer should check back periodically to make sure that the firefighter on the line is okay and that the needs of the backup sector are being met.

The Mission of Extrication

Extrication's mission is to coordinate the removal of victims when mechanical aids must be used. I believe the officer assigned to extrication should not have pneumatic tools in his hand. Just as with attack, search, and the other fire sectors, the officer assigned to extrication should focus on his specific area only. To be effective, he should not be a hands-on guy. He should stand back and ensure the safety of the extrication crew and the victim, as well as determine the area(s) to be opened (cut). There are two forms of extrication—manual and mechanical.

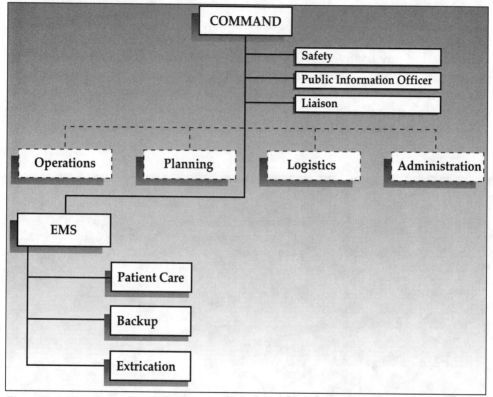

Figure 21-6. The relationship between Extrication and other sectors at a vehicular accident.

Manual Extrication

In a manual extrication, crew members can remove the victim from the vehicle without the aid of tools. Usually we board a victim in the car and then slide the board out of the vehicle and onto a cot or the ground (or an elevated object such as a first-aid box, awaiting the arrival of a cot). No formal sector assignment is required at incidents at which only manual extrication is necessary. Removing the victim is just the step between immobilization and packaging or packaging and transporting in the patient care sector.

Mechanical Extrication[4]

The crew assigned to patient care may need mechanical assistance to

Figure 21-7. An example of manual extrication. (Photo by author.)

remove a victim from the vehicle. Usually the vehicle doors have been damaged and will have to be cut open with a pneumatic tool. Extrication may also have to remove the steering column, roll up the dash, and remove the seat and the roof. Special rescue units usually carry heavy extrication tools. Members of these units must be proficient in the use of these tools and must maintain their skills by training at regular intervals.

The extrication officer is responsible for making sure that the appropriate means of mechanical extrication are used at the appropriate time. There must be control and coordination among the patient care, backup, and extrication sector officers at vehicular accidents

Figure 21-8. An example of mechanical extrication. (Photo by Tom Jaksetic.)

requiring mechanical extrication. All must work together. Since they and their crews often work in close proximity, lines of communication and coordination must be maintained.

The Mission of Traffic Control

Traffic's mission is to ensure that other moving vehicles in the vicinity do not interfere with rescue efforts. I believe that traffic control is a function of law enforcement personnel, not the fire department. Firefighters are not trained to direct and reroute traffic. However, there may be times when members of the law enforcement community are not available to stop or redirect traffic. At these times, apparatus may be used to close down a street. After initially placing the apparatus, the crew can leave it there and go to work in other areas.

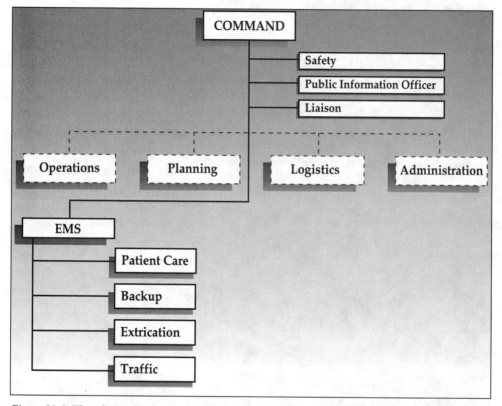

Figure 21-9. The relationship between Traffic Control and the other sectors at a vehicular accident.

If Command must use a firefighter for traffic control, that individual should be given a portable radio. Command should again call for a law enforcement unit (police, sheriff, or security guards, for example) to take over this sector. Explicit directions should be given to the police or other agencies handling this sector. Sometimes the police don't consider it necessary to shut down a lane or an entire street. They may try to talk you out of your intent to reduce or stop traffic. My only caution here is that when you are Command, ultimately you are responsible for the incident's outcome and your crews' safety. If a member is struck by a passing car or truck, and if it can be argued that the member would not have been injured or killed had the traffic been stopped, you may be held liable. (Anytime you are aware that a dangerous condition exists and do not take the measures that a "reasonable, prudent" fire officer might take, you may be liable).

The Mission of Transport Staging

Transport Staging's responsibility is to send the appropriate transport unit to the scene when Command (or another appropriate sector officer) calls for one. This sector is identical to Level II staging at a large fire incident, except that the individual assigned to be Transport Staging need not be an officer. At an incident where more than two transport vehicles are needed, a transport staging sector should be considered.

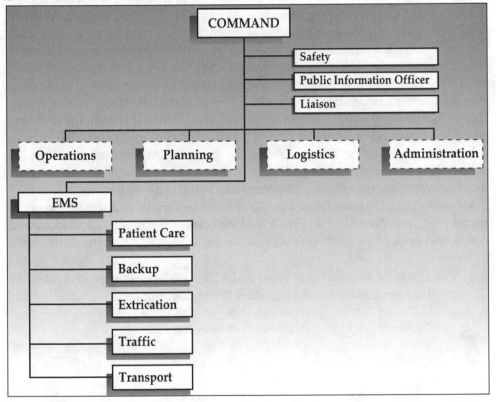

Figure 21-10. The relationship between Transport Staging and other sectors at a vehicular accident.

If Command initiates a transport staging sector, he should notify the entity that dispatches transport vehicles that he has done so and ask that all transport units requested from that point on be sent to the specified staging area. A firefighter with a portable radio should be sent to the staging area to take the role of transport sector officer. The radio designation could simply be "Transport."

When Command requests a transport vehicle, he should specify the location to which it should report—perhaps to a specific vehicle or near a specific engine. Transport may have to make a crude sketch of the area to show the unit where to report. Command must be kept informed of the number of transport vehicles remaining in staging.

At large incidents, a transport officer may be assigned to work with a triage or treatment sector officer. In this case, the transport staging officer should work under the direction of the transport officer. In these large incidents, the entire EMS section of the incident should be on a separate radio channel. The radio designation of the transport staging officer (who works under the transport officer) can simply be "Staging."

GEOGRAPHICAL SECTORS AT VEHICULAR ACCIDENTS

Sometimes the complexity of the incident or its location may make it necessary to divide the incident into more than just functional sectors. At these incidents, georgraphical sectors can be used. The geographical assignments can be similar to those made at structure fires. Probably the most common way to sector an accident geographically—and, I believe, the most difficult—is to use compass directions. When using compass geographical sectoring, two basic assumptions must be made. The first is that the street you're working on is relatively a north-south or an east-west street. It is difficult to determine the direction of winding streets, crooked streets, and those that run on a bias. Second, you're assuming that the person you are assigning knows which direction you're talking about—in essence, which way is north? As far as I know, engine and truck companies are not given compasses.

My solution is to establish all geographical sectors in relation to the command post, as is done at a fire. That's the key. Look at Figure 21-11 and locate the command post. Now, assume you are listening to the following transmission.

Dispatch: Engine 7 is at Front and Main. We have a three-car accident in the intersection. Engine 7 is Main Street Command.

Command to Engine 7: Take the car in Sector C.

Command to Engine 5. Take the car in Sector A.

Command to Engine 6: Take the car in Sector B.

Officers responding to the incident and listening to this transmission (which, by the way, takes approximately 15 seconds) know the following:
- Engine 7 is on the scene and there is a three-car accident,
- Engine 7 officer is Command,
- the command post is located on Main Street at or near Engine 7, and
- the specific responsibility to which each has been assigned.

The type of incident does not change command post assignments. This type of incident would require formal command (more than one unit responding) and a stationary command post. We know that Command has

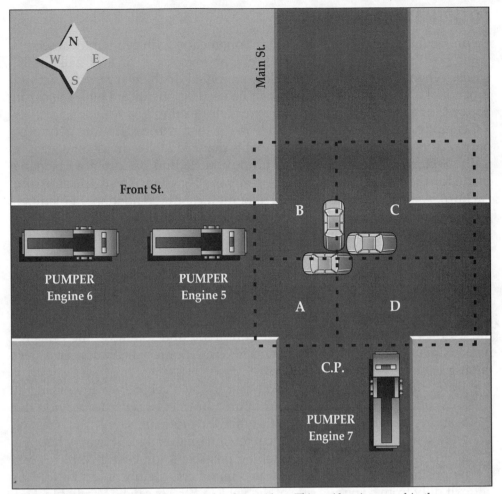

Figure 21-11. Geographical sectors at a vehicular accident. This accident is sectored in the same manner as the inside of a structure.

established "Main Street Command," which means we can assume that Engine 7 and the command post are probably on Main Street. Assuming so, approaching units will be able to see Engine 7 and, without hesitation, should be able to envision a square within which the accident area is enclosed. Determining the area of responsibility should then be easy.

Remember: When only geographical sectors are used at vehicular accidents, an area or vehicle—not a functional sector—is assigned. If I were one of the officers responding to the above accident and Command had just given me the assignment of Sector A, I would have had to assume that I would be responsible for handling everything that happens within that sector.

The type and color of vehicles involved can also be considered geographical sector designations. The red Ford or the green car will suffice as long as only one red or green car is involved in the accident.

COMBINATION SECTORS

Functional sectors should be used in conjunction wtih geographical sectors when possible. They help define the roles and responsibilities of those at the scene. When more than one vehicle is involved or there are many victims, functional and geographic sectors can be used—patient care with the car in Sector C or extrication in the pickup truck, for example.

Generally, patient care—and, to a lesser extent, extrication—are about the only sectors at typical vehicular accidents that need to be divided geographically. Sector assignments such as traffic control and backup are normally made for the entire scene (only one company is assigned to extrication and backup). As such, they do not need geographical differentiation. Remember, if Command simply designates an officer to be backup, that officer's responsibility is to protect all the victims and personnel from flash fires. The line must be positioned in an appropriate location.

SECTOR OFFICER RESPONSIBILITIES

The responsibilities of sector officers at vehicular accidents are identical to those of sector officers at a structure fire. First and foremost, the officer's responsibility is to protect himself and his crew. Some specific concerns pertaining to crew safety are as follows:

- Traffic. If a crew member should be struck by a vehicle while working at an accident, we will flock to our injured brother or sister to provide the best care possible. We will do this while the original victims (the ones we were sent to help in the first place) remain unattended. This may happen at times, but we must do everything possible to prevent it. Vehicles can be positioned at an angle to protect members. Members can be told to exit only from a specific side of the vehicle. Streets can be totally blocked with an engine or truck. Police can be called on the scene to stop, redirect, or slow traffic.
- Flash fires. If your hands are full of victims and you smell gasoline, notify Command and tell him that you think a backup sector needs to be established. Perhaps Command hasn't noticed the gasoline or other fluids on the street. Remember, though, that if Command assigns you to patient care, he does not expect you to be pulling backup lines. He expects you to have collars and blood-pressure cuffs in your hand. If no lines are pulled and one is needed, tell Command.
- Bloodborne pathogen protection. Make sure your crew is "gloved up." We forget sometimes. This is especially true of new members and drivers who are rarely required to help with victim care. (In some cities, drivers are required to glove up.)

The second responsibility of the sector officer is to *complete the task assigned*. Focus on your assigned activity. Don't freelance. If you see something else that needs to be done, inform Command.

The sector officer's third responsibility is to *keep Command informed*. Use benchmarks to let Command know that your task has been completed. Let Command know if something unexpected occurs and if you see something else that needs to be done.

BENCHMARKS FOR VEHICULAR ACCIDENT SECTOR ASSIGNMENTS

Patient Care

A specific benchmark for patient care may not be required. If there are few victims and the levels of treatment and packaging are minor, a benchmark may not be needed. If the accident is major and there are many victims and geographical sectors, a benchmark may be necessary so that Command or Operations will be aware that the victims have been extricated, packaged, and transported (or prepared for transport).[5] The benchmark "Victims treated" may be used.

Backup

When hazardous fluids are evident or mechanical extrication is necessary, a backup sector (with a charged line) should be used. As with firefighting operations, the commonly used benchmark for backup is "Backup line in place." When Command receives this benchmark, he knows that a line has been pulled and charged and that a firefighter is on it. (Extrication may have to wait until he hears or is informed that the backup line is in place).

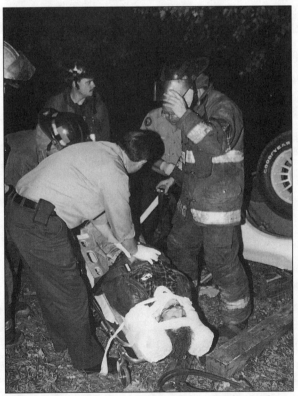

Figure 21-12. This victim has been packaged and turned over to transport crews.

Extrication

This benchmark is generally required only when mechanical extrication is required. Once the victims have been removed from the vehicle and are being "boarded," Extrication gives the benchmark "Victims extricated." At this time, if only one extrication company is on the scene, Command or Operations can assign extrication to another vehicle, reassign the unit to another sector (such as treatment), or tell it to start to pick up.

Traffic Control

If fire personnel have been assigned to shut down the street, Traffic Control will give Command the benchmark "Traffic stopped" after the task has been completed. Command may relay this information to crew members working in the area, who will be able to move about a bit more freely without the risk of getting struck by a car.

Transport Staging

After all the victims who were to be transported have been taken from the scene, the benchmark "All victims tranported" is given. Command then understands that all of the victims have been triaged, treated, packaged, and transported. Transport agencies and units listening will understand that they probably will not be needed at the scene.

These benchmarks are recommendations that can be used at vehicular accidents. If they don't fit your operations, that's fine. I believe in the concept of benchmarks, however. When I am Command, I don't have to hover over anyone's shoulders to know when things are being done. I can focus my attentions elsewhere, knowing that I will be "fed" the information I need.

Endnotes

1. One firefighter (EMT) for C-spine immobilization and the other for taking vital signs and administering patient care.

2. This is a minimum for patient stablization. If intubation and IV lines are needed on savable victims, more EMTs and medics will be required.

3. Rescuer-victim ratio of two to one.

4. Mechanical extrication is commonly referred to as "heavy extrication." They both mean the same thing.

5. I recommend that no victim be left alone while waiting transporting. If the victim can walk or if several victims are in a specific area, a firefighter should be left with them. Some may consider this a waste of personnel, but if the patient goes downhill, a lawsuit may be possible. The grounds will be "patient abandonment." You have a duty to ensure the safety of wounded victims while they are on the scene. Once the victim is in the back of a transport vehicle, the responsibility may shift. Consult your legal department for advice on this subject if procedures are not already in place in your department.

REVIEW QUESTIONS

1. List the points that should be covered in a conditions report at a vehicular accident.

2. What are the guidelines for selecting a Level I staging location at a vehicular accident?

3. Transport staging works in the same manner as _____.

4. Who should conduct the initial triage at a vehicular accident?

5. What three types of sectors are used at a vehicular accident?

6. Give an example of each of the types of sectors used at vehicular accidents.

7. What sector is most frequently assigned at a vehicular accident?

8. What are the four components of patient care?

9. What is the minimum size backup line that should be pulled?

10. Should the backup line be charged? ___ Yes ___ No

11. What are the two forms of extrication?

12. Who should be responsible for traffic control at a vehicular accident?

13. Why are combination sectors useful at incidents?

14. List the three responsibilities of sector officers at vehicular accidents.

15. What three hazards jeopardize crew members at a vehicular accident?

DISCUSSION QUESTIONS

1. Why do you believe the author stresses "minimums" for rescuer/patient ratios on EMS runs? Do you believe this to be just a "Toledo thing," or should they be established in other departments. Why?

2. Describe your department's "traditional" method of assigning sector responsibilities at vehicular accidents. Compare and contrast your way with the author's way.

3. Compare and contrast backup at a fire with backup at a vehicular accident.

22

Hazardous-Materials Incidents

TOPICS

- The three levels of hazardous-materials incidents
- The levels of hazardous-materials incident response
- The responsibilities of Command at hazardous-materials incidents
- The duties of the first-arriving unit at a hazardous-materials incident
- The command structure at a hazardous-materials incident
- The responsibilities of Scene Safety at a hazardous-materials incident
- Police functions at a hazardous-materials incident
- The functions of Command at a hazardous-materials incident
- Location of the command post at a hazardous-materials incident

KEY CONCEPTS

- ✓ Cold zone
- ✓ Decon
- ✓ Demobilization
- ✓ Entry suits
- ✓ EPA (Environmental Protection Agency)
- ✓ Haz-mat officer
- ✓ Haz-mat safety officer
- ✓ Hot zone
- ✓ Jurisdiction having authority
- ✓ Level A suits
- ✓ Level 1 incidents
- ✓ Level 2 incidents
- ✓ Level 3 incidents
- ✓ Perimeter
- ✓ SARA Title III
- ✓ Warm zone

Managing hazardous-materials incidents poses some specific problems for fire personnel. Several factors will influence how the incident will be structured and mitigated. As discussed in Chapter 2, SARA Title III states that the jurisdiction having authority responding to a hazardous-materials incident shall operate under the incident command system (ICS). This applies to all fire departments—even those that do not own a nonsparking shovel, let alone Level A suits—as well as hazardous-materials teams. If you respond to a hazardous-materials incident, you must operate under the ICS. If trucks travel through your protection area or if a set of usable train tracks is within your jurisdiction, the potential for a hazardous materials incident exists.

TYPES OF INCIDENTS

There are three types of hazardous-materials incidents:

- The first, a *Level 1 incident*, is defined as an incident "that can be handled by the first responders at the operations level, with no 'outside' or greater-level-trained intervention."[1] The most common type of Level A incident in my jurisdiction is a small flammable-liquid spill that usually occurs at or near gasoline filling stations or at small industrial areas.

- The *Level 2 incident* covers small but bona fide hazardous-materials spills that "can be handled with the aid of haz-mat technicians, but normally are relatively small in scope and do not transcend jurisdictions or outside entities."[2] This level of incident would include leaking 55-gallon drums in the back of semitractor trailer rigs and minor ruptures of piping and fittings at industrial sites. These incidents are a bit more complicated than Level 1 incidents. Level A suits or flash suits and nonsparking tools may be required. Sometimes minor evacuations may be necessary.

- *Level 3 incidents* include "the big ones." A Level 3 incident is defined as one "involving a severe hazard or a large area that poses an extreme threat to life and property and that may require a large-scale protective action."[3] These incidents may require resources beyond those available in the community and can tax even the largest departments. Standing back and observing Level 3 incidents is like watching a three-ring circus. Numerous activities are taking place: Entry teams are suiting up, decon is being set up, evacuation is underway, and medical sectors for haz-mat responders are being established.

A strong incident management system is needed at Level 2 and Level 3 incident response operations; that is the only way to ensure a successful outcome. Lines of authority must be established and sectors (responsibilities) defined. If Command can delegate the resolution of the problems he detects or that have been presented to him, the incident will go smoothly. However, if freelancing is allowed, the operation is not adequately controlled, and communication among the various responding agencies is lacking, the incident could prove to be overwhelming and become a disaster.

LEVELS OF RESPONSE

An engine company is the minimum typical response at a Level 1 incident. The company is sent to check when there is a spill of a fuel or other hazardous material. Such spills often occur on private property.

A review of the basics: A one-unit response means the operation will be under informal command. The incident type doesn't change this—when one unit responds, informal command is established. The officer is responsible for the entire operation.

You don't need a third alarm for a five-gallon gasoline spill at a Jiffy Mart filling station. You need common sense—turnouts on and a hoseline pulled or an extinguisher out with the pin removed if the spill is small. Eliminate potential sources of ignition. If the leak is not already stopped, stop it if possible. Cover the spill with inert, nonsparking materials. Notify the local pollution control center or Environmental Protection Agency and the sewer department if the material enters the sewer system.

Finally, remind the owner/occupant of his responsibilities. The property owner is responsible for having the spilled material "properly" cleaned up and "properly" disposed of. We usually tell the owner where in the phone book to look for contractors who do this type of work; we do not recommend specific companies. The Environmental Protection Agency also will provide the owner with information.

At this incident, Command covers it all single-handedly. He is Command, Liaison, PIO, and Safety, as well as Operations, Planning, Logistics, and Administration.

Level 2 haz-mat incidents necessitate that more tasks be done. Consequently, the response increases. For incidents involving leaking drums on tractor-trailer rigs or a loading dock, for example, up to a regular- or first-alarm response, plus a haz-mat unit, may be required. In smaller communities, this type of response usually is not possible for a "minor" spill. The department should send what it can.

With this type (or size) of response, the officer of the first unit to arrive on the scene takes command. All other responding units stage.

The same type of response is needed as for major incidents. In fact, many major incidents start out minor. Once the first unit arrives, formal command must be established.

PRIORITIES AND RESPONSIBILITIES OF COMMAND

When the first unit arrives at a haz-mat incident and establishes command, Command must first look at the picture and review the four priorities of command, which are reviewed in the following account of a hazardous-materials incident to which I responded in February 1996.

INCIDENT

The incident occurred at a major hospital that had a new attached wing that housed a physical fitness-therapy health-care facility. The call came in around 0800 hours on a weekday. There was a leak near the whirlpool area in the basement of the wing. While walking to my car, I was thinking "chlorine."

I was at the scene within two minutes. Unit 136, our safety officer, arrived about 10 seconds before and reported the following:

Unit 136 at 1601 Superior; 136 is Superior Command. An evacuation is in progress.

I pulled into the rear of the parking lot and motioned that I was taking command from him.

Battalion 1 is Superior Command.

Firefighter Safety

My first concern was for my crews. I had to determine whether we could enter and operate, and at what level. I motioned to a security guard to come over to my car. Ever since my confined-space training, I follow the new practice of having people come to me at this type of incident. The guard told me that a "feed line" to the whirlpools had burst and that about 10 gallons of liquid chlorine spilled on the floor. To ensure firefighter safety, I radioed this information to the haz-mat unit that was responding. I asked what level of personal protection personnel needed and whether crews could go inside and search for and remove employees. If I hadn't taken this step, I wouldn't have been able to go to my second priority (civilian safety) for quite a while.

It was determined that crew members could enter in Level D suits and with their SCBAs in the standby position. If they detected odors, they were to mask up and exit the building.

Civilian Safety

With crews in the appropriate level of protection, we searched the structure to ensure that all occupants were out. We then established a perimeter with scene tape and stationed crew members at that site so that no one could gain access to the building. Several security guards and the maintenance man on duty "took in" small amounts of chlorine and were experiencing initial symptoms. Our medics evaluated them and then took them to the hospital emergency room for treatment.

Stop the Problem

Now that our members could enter the structure to ensure the safety of civilians, my next concern was to stop the problem. I approached this task slowly, methodically, and with "control." I sat in my Jeep® and just waited. The building was not occupied, and no one could enter. The remainder of my crew were outside, standing down while the haz-mat team was preparing to enter the structure and assess the situation. Two haz-mat members in Level A suits entered and surveyed the scene. They determined that the flow of the liquid chlorine had stopped and that approximately three gallons of the liquid were on the floor. Readings were taken and recorded. The HVAC system was used to remove the fumes. We took several more air samples and determined that the spill had been stopped. Shortly thereafter, readings showed that the level of detectable residual fumes was acceptable. The haz-mat team was decontaminated and then began to pick up.

Conserve Property

We were careful to ensure that damage to the building and its contents would be minimal. Doors were closed, and ventilation was begun as soon as it was practicable.

As Command, my task at this incident was basically to sit back and let my crews handle the situation. I supported my members by ensuring that their safety was paramount in my mind. The first thing I did on arrival (after giving a brief on-scene report) was to determine the appropriate level of protection for crew members and whether they could enter the structure to search for victims. They knew their safety was my priority.

THE DUTIES OF THE FIRST-IN UNIT

We have established that the first unit on the scene of a haz-mat incident should establish command. What can Command expect of personnel who are not haz-mat technicians at a minor or major haz-mat incident?

Command needs to concern himself with three things while the haz-mat unit is responding or suiting up. In order of importance, they are as follows:

- *Identification.* Attempt to identify the hazardous material involved. Unless Command is the only responder on the scene, he should not be looking through binoculars. That's not his job. As you know by now, if Command is looking through binoculars, he can't be focusing on the whole incident.

Figure 22-1. The placard on this tanker indicates the contents. If the first unit on the scene can identify the material, it has done its job. (Photo by author.)

- Recall that in Chapter 3 we discussed what Command should do with the remainder of his crew when he takes command. In this instance, Command would give one member a pair of binoculars and have him try to identify the material. Except for extreme cases, once a first responder has identifed the spilled material, his job has been completed. It's the most important task a first responder can, and should, do.
- *Make any "feasible" rescues.* People may be lying all over the street. Stop and ask yourself why. Prior to any rescue, *identify*. Don't become part of the problem. Tell the responding haz-mat unit what the placard number is, and paint him a picture of the scene (if he's still en route). Then ask him two questions:
  —Can rescues be made?
  —What level of clothing is needed?

If you come on the following scenario, your instincts may tell you to act differently.

SCENARIO

A truck driver is conscious but unable to get out of his truck. He's waving his arms and asking for help. You think, "If this stuff hasn't killed him by now, we can mask up, run up, and pull him out of the rig."

Before you do this, remember that some hazardous materials rolling down your roads every day may take days, weeks, or even years to kill. The driver in the above scenario may be alive, but his "clock" may have started ticking. Resist the temptation to run up to the truck. Check with the haz-mat team first. Once the team confirms that it is safe to make a rescue, do it; then begin to triage and treat victims.

- *Start evacuations.* Refer to the Department of Transportation (DOT) *Emergency Response Guidebook* (DOTP5800-1994) or other reference books carried on your rig, and ask the responding haz-mat unit whether evacuation is necessary. If no evacuations are needed (the spill is already isolated by considerable distance or is too small for an "official" evacuation), start to establish zones for the incident.

Again, Command should not be identifying, rescuing, or evacuating. These sectors should be assigned to other responding units.

COMMAND STRUCTURE

The command structure at some haz-mat incidents can be as simple as that illustrated in Figure 22-2. It would be ample to handle the gasoline spill at the Jiffy Mart discussed above. The type of command would be informal. Areas denoted by dotted lines indicate responsibilities or roles Command has chosen to handle himself. A safety sector officer and a spill control officer (who covered the spill with sand) were designated. It's usually not necessary to diagram an incident of this type. The flowchart is presented here to

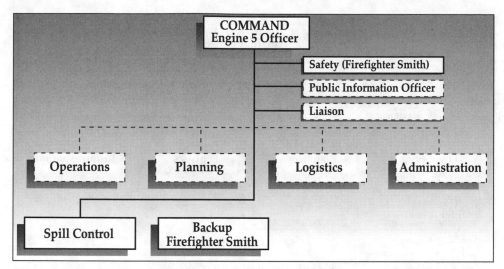

Figure 22-2. This flowchart applies to a small "assist" haz-mat involving a gasoline spill at a gas station.

illustrate the relationship between Command and his crew and to show that Command is responsible for certain things at every incident. I would not expect Command (Engine 5 officer) to make these assignments over the radio at an incident like this. However, I believe that anyone who asks the officer of Engine 5 for a report on this incident would have to be impressed if such a diagram were attached to his report. I know I would be; it would show me that he is well acquanted with the components of incident management.

I was Command at the incident on pages 339-341. When the haz-mat unit arrived, I assigned the senior lieutenant to be the haz-mat officer. I was then better able to give my attention to other aspects of the incident. I was responsible for all the staff functions designated by dotted lines in Figure 22-3. The haz-mat officer, on the other hand, was directly in charge of the majority of the personnel who responded. He handled the spill. I made sure that everything that needed to be done was being done. He focused on the spill without having to worry about the media, police, or maintaining the perimeter. I didn't have to think about decon, entry, or the level of personal protective equipment required for the entry team. Once the haz-mat officer decided on a mitigation plan, determined the level of protective clothing required, and selected the location for decon, we met to discuss our options. I then gave him the go-ahead. The system works because each sector focuses on its own tasks.

SECTOR ASSIGNMENTS

Some specific sector assignments must be made at a haz-mat incident. Most of them fall under the jurisdiction of the haz-mat officer, who is under Operations. These sector assignments are as follows:

- *Haz-mat safety officer.* Responsible for the safety of the entry team, he helps determine the level of protective clothing needed and the supplies required for decon. He works directly with the haz-mat officer, but in

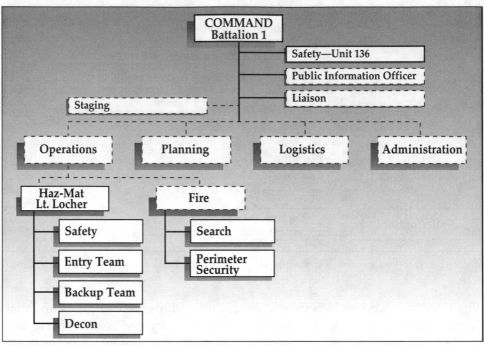

Figure 22-3. This is the flowchart for the chlorine spill previously discussed. Notice that a haz-mat offi- cer was assigned to handle the haz-mat portion of this incident. Command was responsible for all of the staff roles and functions depicted by dotted lines.

reality he works for the scene safety officer. This officer must have haz-mat training at least to the technician level.

- *Entry team.* Its members are the haz-mat technicians who will don the required suits, turn the valves, and plug the holes. Personnel assigned to this sector require specific training (which will not be covered here).

Figure 22-4. A chief officer takes the role of a sector officer at a plane crash involving hazardous materials.

- *Backup*. This team is made up of trained haz-mat technicians who are prepared to enter and aid the entry team if it should encounter trouble.
- *Decon*. This sector decontaminates personnel, tools, and sometimes victims. Detailed coverage of entry and decon are beyond the scope of this text.
- *Medical unit*. This sector must be established when an entry team dresses. The members should be paramedics and must focus on the health and physical status of the haz-mat team members. They must monitor and record the team members' vital signs.

Command is responsible for operations at the incident depicted in Figure 22-7. He has divided this function into three sectors. This is a haz-mat incident. As such, the fire department must provide some type of haz-mat service, whether it be done by crews from within the department or by outside technicians. In addition, the department

Figure 22-5. Decon procedures for fire personnel involved in a lubrication oil fire at an abandoned heat-treating plant.

may have to address other needs that may have been created by the initial haz-mat incident, such as EMS and fire suppression.

THE FIRE AND EMS SECTORS

Fire and EMS sectors usually are established at Level 3 (sometimes at Level 2) incidents and are under the Operations officer. These sectors are separate and distinct.

The Fire Sector

At Level 2 and Level 3 incidents, Command must focus on matters other than the spill. Haz-mat incidents can be dynamic and multifaceted. I believe that's the reason the fire department, and not other emergency response agencies, is sent to haz-mat spills. Specific operations or tasks can be conducted under a fire sector at a haz-mat incident. The threat of fire is present at many haz-mat incidents. even when there is no fire and the material spilled is nonflammable, the fire department may have to use water streams to disperse vapors.

If there is an ensuing fire at a haz-mat incident, lines may have to be pulled to protect exposures and *possibly* extinguish the fire. It may be best to let some materials burn if they become involved in fire. Always check with the haz-mat team or sector before throwing water at a haz-mat spill. If the potential for ignition exists, unmanned lines may have to be set up.

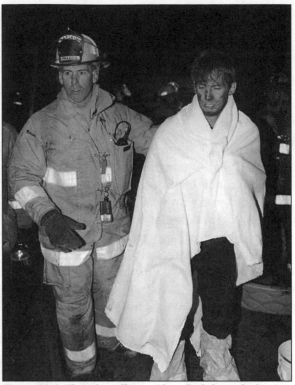

Figure 22-6. The safety officer guides a firefighter who has just been decontaminated.

Search

At times, crews may be required to search for and remove victims trapped or otherwise affected by a hazardous-materials spill. After the haz-mat safety officer has determined the appropriate level of protection, fire crews *qualified to don that specific level of protection* can be used to search for and rescue victims.

Evacuation

After consultation with the haz-mat officer, proper evacuation distances, which may range from a few hundred feet to several miles, may be established. I can envision the whys and whens of evacuation. I am baffled by the hows. I teach hazardous-materials response at a community college in the Toledo area. My favorite section of the course, and the one that draws the most discussion, is evacuation. Whether you provide protection to a rural or urban area, the concern is how to evacuate a two-square-mile area safely and in a timely fashion. My only thoughts are "police" and "media." You will have to rely on the police to help in evacuation. They should start at the extreme portions of the evacuation areas and work toward the spill. Use them in areas that do not require the wearing of protective clothing. The media can broadcast warnings over the radio and television. Provide them with detailed instructions including the locations of emergency shelters.

Perimeter

The police are better equipped to handle this aspect of an incident, but they aren't always available.

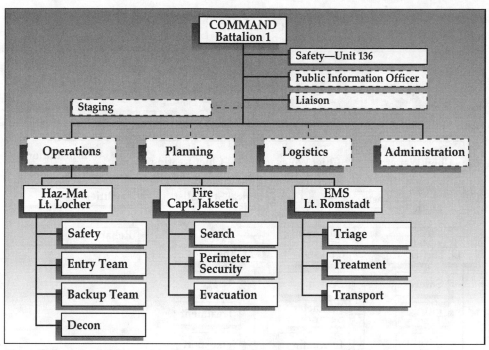

Figure 22-7. A Level 2 haz-mat incident. Note that Command has not given up the operations function at this incident.

Emergency Medical Sector

At some incidents, an emergency medical sector for injured civilians must be established. Certainly hundreds of chemicals out there have the potential of injuring anyone exposed to them. At some incidents, civilians may have been injured by the accident that precipitated the spill. At these incidents, an EMS section is needed to triage, treat, and transport the injured and exposed victims. At incidents that necessitate an EMS sector, the following probably will have to be established at the minimum.

- *Triage.* After fire personnel assigned to search have brought out the exposed victims, the victims must be triaged to determine who should be treated and transported first. Some communities have agreements with hospitals to have emergency room physicians respond to mass-casualty incidents to perform triage and administer treatment.

- *Treatment.* After the victims have been triaged, those needing minor treatment may be tended to while awaiting transporting. Victims who suffered trauma will have to be packaged. Others may need only to be supported with oxygen and have their vitals monitored. It may be necessary to establish a field hospital for treating victims at the scene. Necessity will dictate the level of on-scene treatment to be provided.

- *Transport.* Victims at the top of the transport list will be taken to a health-care facility. Depending on the situation, a transport staging sector may be needed to coordinate the transport vehicles in the area.

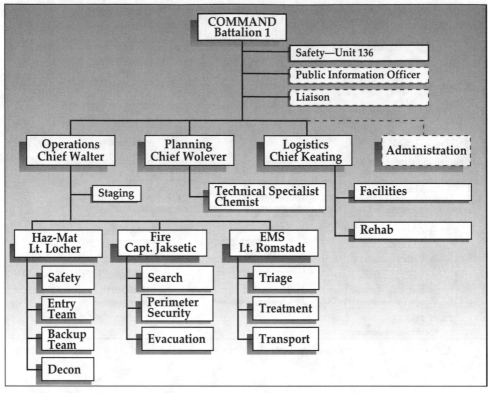

Figure 22-8. Command structure at a Level 3 haz-mat incident.

Level 3 hazardous materials incidents are among the most taxing incidents to command. Although I do not cover these types of incidents in this text, I have prepared a diagram of such an incident, which is presented in Figure 22-8.

At this incident, Command has only three people, not including the safety officer, directly reporting to him—the operations, planning, and logistics officers. As many as 60 or more personnel could be at this type of incident. Command should not be overtaxed working within this command structure. If he should feel overburdened, however, he could delegate additional responsibilities such as public information or liaison.

Safety Sector[4]

Safety, one of the command *staff* roles, is an essential sector at a haz-mat incident. The safety officer is responsible for *scene* safety. He needs to ensure the safety of civilians and support personnel (fire and EMS sectors). As is the case in all incidents, safety is best served when the officer makes his own to-do list (after consultation with Command) and then moves throughout the area (excluding the warm and hot zones) to assess each area of concern. A haz-mat scene must be taped off. Police or other security officers can help maintain the perimeter. The air must be monitored, and Safety should receive frequent updates on weather conditions.

USING THE POLICE

Sometimes it may be necessary to use the police at a haz-mat incident. They are trained in how to keep traffic flowing and will attempt to do it even when it may be counterproductive to the fire department's interests. I have been at incidents at which the police insisted on keeping the traffic moving in lanes immediately adjacent to areas in which crews were working. When Command orders a street to be shut down, traffic problems increase, creating difficulties for the police. When a major street is closed to traffic, the police find themselves in a situation comparable to that experienced by the fire department when a second alarm must be called.

Nevertheless, if Command feels that the traffic poses a hazard for civilians or personnel, he must shut down the street. In a situation like this, approach is everything. Ask first. Don't demand. Explain the reason for the action, and most police officers will be glad to help. If the police are not on the scene yet, block the street with apparatus until they arrive. As a last resort, if the police are reluctant to shut down the street, pull a hoseline. In most states, it is against state law to drive (or allow someone to drive) over fire hose. I always advocate the pulling of a protective line. Once pulled, it's hard for an officer to justify allowing traffic to cross over the line, and who is he to argue the need for a hoseline with a fire chief?

In summation, the police should be considered for the following tasks at a haz-mat incident:

- *Traffic control.* Traffic into the spill area must be restricted at a large incident. Routes may have to be diverted, and barricades may need to be erected.
- *Evacuation.* Let the police start at the most remote sections of the area to be evacuated. Discuss this with Police Command (if there is such a thing), your haz-mat safety officer, the haz-mat officer, and Command or Operations. If the risk is low, they can cover much of the area working back toward the spill.

COMMAND'S FUNCTIONS

The planning officer may play a very crucial role at a large haz-mat incident. Command and the operations officer brief the planning officer concerning the type of material and quantity involved, the exposures, and other evacuation concerns. If technical expertise is needed, it must be funneled through the planning sector. Planning will meet with and glean information relevant to the situation from technical specialists called to the scene. Some departments have contracted with private-sector chemists to be available for large or complicated spills on an on-call basis. A chemist can provide valuable information, such as what will happen when chemical A mixes with chemical B. After scenarios have been created and courses of action have been decided, Planning should inform Command and Operations of the options. At this time, Logistics may need to get the necessary supplies, equipment, and services to the scene. Finally, if an Administration Sector has been established, that officer will be consulted to determine who will pay for the services and supplies and where the bill should be sent.

Look at all the things with which Command does not have to concern himself. He serves as the final "sounding board" and the individual who puts it all together.

Planning will also be responsible for de-escalation, referred to as *demobilization*, of the incident. At a Level 3 incident, scores of units may be on the scene. Once the incident is declared to be "under control," the scene will have to be picked up in a logical manner. Planning may also be placed in charge of notifications, which consists of establishing a list of who must be contacted if a large amount of a hazardous substance is released. Some departments have preestablished lists for specific types of incidents. Such a list can be a timesaver. It's better to sit in an office brainstorming with peers about who should be included on a resource list before an incident occurs than to be in the command post trying to come up with names out of the phone book under the pressures of an incident in progress. Be prepared. Remember: You have to look good to be good.

COMMAND POST LOCATION

Selecting the site for the command post requires special attention at Level 2 and Level 3 haz-mat incidents. Some guidelines, including the following, should be kept in mind:

- The command post should be out of harm's way, upwind from the spill. Check the *DOT Guidebook* or other reference sources for the proper distances for zones.
- It should be on the fringe of the Cold Zone, just beyond the scene tape that separates the Warm Zone from the Cold Zone.
- It should be slightly remote from but within easy walking distance of the haz-mat unit. The reason for this is that there is a great temptation to micromanage the activities of the haz-mat unit if it is immediately adjacent to the command post. Let the team members do what they were trained to do. Don't watch them suit up, and, in turn, they won't interrupt your discussions with officers and representatives of other responding agencies. When courses of action have been determined, the Haz-Mat Sector Officer and Command or Operations can confer.

Endnotes
1. Noll, Gregory G., Michael S. Hildebrand, and James G. Yvorra. *Hazardous Materials: Managing the Incident,* 2nd edition. 1995. Fire Protection Publications, Oklahoma State University; Stillwater, Oklahoma.
2. Ibid.
3. Ibid.
4. The safety sector referred to here is the scene safety officer, not the haz-mat safety officer. The scene safety officer is one of Command's staff positions, as discussed in Chapter 4.

REVIEW QUESTIONS

1. Define the "jurisdiction having authority."

2. What are the three levels of haz-mat incidents?

3. What are the three duties of the first-arriving unit on the scene of a haz-mat incident?

4. What is the most important duty of the first-arriving unit?

5. List the specific functional sectors associated with a haz-mat incident.

6. List the three sectors that will normally be filled under Operations at a Level 3 haz-mat incident.

7. What four sectors fall under "Fire" at a haz-mat incident?

8. In what two areas can the police be of assistance at a haz-mat incident?

9. Under what section will chemists work at a haz-mat incident?

10. Where should the command post be established at a haz-mat incident?

DISCUSSION QUESTIONS

1. Compare and contrast the duties of the two safety officers at a haz-mat incident.

2. Discuss rescues at a haz-mat incident prior to identification. Assume you are the chief of a department and are asked to justify to the city council and the mayor the reason no rescues were made at a haz-mat incident to which your department responded. At the incident, two victims died while onlookers watched your department stand back. Justify the fire department's decision.

3. The author discussed pulling lines if the police fail or refuse to stop traffic at a haz-mat incident. Cite specific incidents at which you had a problem with the police and what you or your department did to resolve it.

23

Miscellaneous Incidents

TOPICS

- The types of command used at alarm system responses
- The significance of firefighter safety and civilian safety at a water rescue incident
- The command structure at a water rescue incident
- The responsibilities of the dive master
- The functional sectors used at a water rescue incident
- The basic types of bomb incidents
- Who should serve as Command at each type of bomb incident
- When firefighters can enter a structure that may have an unexploded bomb
- Using radios at a bombing incident
- Police investigation as part of the command structure at an incident involving a bombing

---

As stated throughout this text, the incident management system should be used at all incidents to which a fire department responds. We have discussed its application to residential structure fires and, briefly, to larger fires; to accidents; and to haz-mat incidents. We now will look at its application to several miscellaneous types of incidents that are common to most departments.

KEY CONCEPTS

✓ Dive master
✓ Exposure suits
✓ Personal flotation device (PFD)
✓ Still box

ALARM SYSTEM RESPONSES

A large percentage of the runs made by fire departments are to check alarm systems in commercial occupancies. Recently, however, there has been an increase in calls for residential fire alarm systems and for carbon monoxide detectors. In my department, these responses are considered "still-box" alarms.[1] Up until recently, we would send an engine and a truck to an alarm-system response in commercial buildings and a single engine to responses in residences. In March 1996, as a result of the increased number of responses citywide, we downsized our response to commercial fire alarms to a single engine company, the same response as for residential alarm calls. Informal command is used at these responses.

With informal command, the officer and his crew enter the structure and determine whether a fire exists (if a fire alarm is sound) or a buildup of carbon monoxide is present (if a detector has activated). Procedure should dictate when entry should be forced if no one is at home at the time of the fire department's arrival. Some alarm companies send representatives in automobiles to respond with a key if the resident is not at home.

Most cities still send a minimum of an engine and a truck to alarm-system responses at commercial structures. Since two units are responding, these incidents are managed under formal command. The first unit on the scene gives an on-scene report that includes the following:

- the fact that it is on the scene and a check of the address,
- a conditions report, and
- the required command mode.

Dispatch: Engine 6 is at 1009 Main Street. Nothing showing.

At this point, units en route know the following:

- there is a fire unit on the scene, and
- there is no evidence of a fire or other emergency at this time.

Since the incident management system is in effect, the first unit on the scene is Command (whether formal or informal command is used), and all other responding units should stage on arrival. If the officer of the first unit on the scene (Command) enters the structure and needs help looking for the source of the problem, he should call a staged unit (if there is one) to help. (Notice that since the IMS has been instituted, someone is always in charge.)

WATER-RELATED EMERGENCIES

Because emotions tend to run high at a water-related emergency, it is vital that the first unit on the scene establish a strong command presence at the beginning of the incident. In addition, the first two priorities of Command must be reviewed immediately on arrival to ensure that the incident will be handled efficiently.

Firefighter Safety

At water incidents, a tendency of the first-arriving crew members may be to jump in and attempt a rescue. Crew members must don appropriate safety gear prior to entering the water. Even the strongest swimmers may be no match for a rapid current or a struggling, frightened victim. The temperature of the water and the speed with which it is moving must be considered as well. Appropriate ropes and personal flotation devices (PFDs) must be used.

Some of the procedures used at water incidents are similar to those used at haz-mat incidents. For example, perimters (hot, warm, and cold zones) must be established, and only members with the appropriate protective equipment (PFD or exposure suits) should be allowed in the Hot Zone. Procedure and training determine the tasks first-arriving members will be allowed to perform. Again, we must be part of the solution, not part of the problem.

Civilian Safety

Here, the term "civilian safety" applies more to onlookers and civilian rescuers than to the initial victim. Scene control is imperative. Again, zone perimeters must be established, and only personnel with appropriate protective equipment should be allowed in the Hot Zone (this applies to members of the media as well). A bad situation can be made worse—if a civilian were to fall in the water, for example.

It's difficult to imagine that there is a jurisdiction that may never have to respond to an incident involving water. Oceans, lakes, ponds, pools, flash floods—all present the potential for a water rescue. At these incidents, a specific command structure, similar to that used at a haz-mat incident, is required. Command must focus on the whole incident and resist micromanaging it.

The Response

Procedure and the availability of specialized units determine the response level at water rescues. My department sends the closest engine, the water rescue unit (with the heavy rescue squad), two medic units, and a battalion chief on all water rescue calls. On arrival, as is the case for any other incident as well, Command can upgrade or downgrade the response. A successful dive operation takes many support units and specially trained crew members.

Command Structure

The command structure for a water-rescue incident resembles that for a haz-mat incident, as discussed in the previous chapter. A dive master usually oversees the dive-team aspects of the incident.[2] The medical unit, dive support, and backup sectors all focus on the rescue diver and his needs. The rescue diver focuses on the dive and on conducting the search, knowing that he has this team looking out for him and his needs. The dive master in turn can focus on the dive to ensure that every technical aspect of the dive is addressed.

Command can focus on the entire incident. He has one individual looking

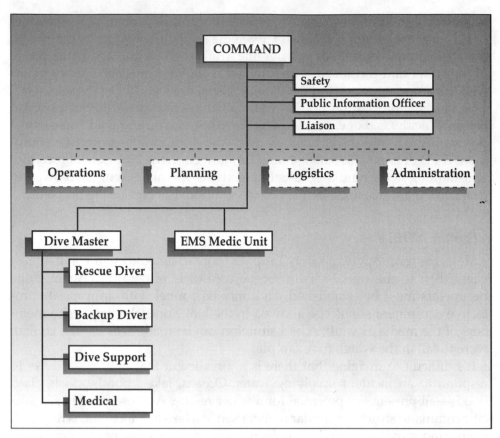

Figure 23-1. A typical command flowchart for a small dive.

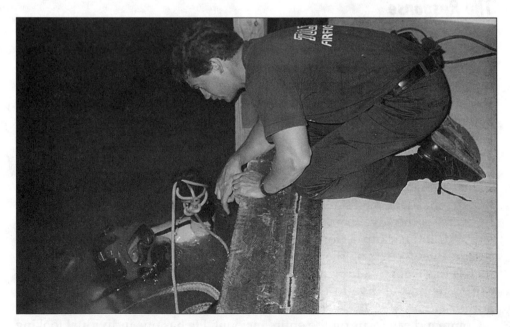

Figure 23-2. The dive master gives last-minute instructions to the rescue diver. (Photo by Ken Anello.)

out for the specific technical aspect of the incident. The remainder of this incident is Command's responsibility, as indicated by the sections and staff positions in Figure 23-1. If I respond to a water-rescue incident, I use a safety officer for scene safety (as previously noted, Toledo has one 24-hour safety officer on duty). I delegate to my safety officer the area of civilian safety, which is one of my priorities.

Figure 23-3. Divers prepare for a quick rescue at hydroplane races.

The following fictitious incident illustrates in detail the command structure presented in Figure 23-1 and how this sytem works at an open-water rescue.

SCENARIO

We are dispatched to a water-rescue incident at 1347 hours on a Saturday afternoon in July. The responding units are Engine 6, Engine 5, Squad 1 with the water-rescue unit, Medic 1 and Medic 2, the fire boat, and Battalion 1.

The caller tells the 911 operator that a boat has overturned under the King Bridge on the Maumee River and that several people are thrashing about in the water.

Engine 6 is at the King Bridge. We have an overturned boat with several people in the water. Engine 6 is River Command.

One of the people in the water swims to shore and tells Command that four people were in the boat, two are still holding on to the boat, and the other man is missing. Command relays this information to the water-rescue unit and the fire boat responding. He tells the water-rescue unit to suit up and orders the second medic unit dispatched (Medic 2) to assess the man who swam to shore. At about this time, a private boat picks up the two other men who were holding on to the boat. This leaves one man unaccounted for. Battalion 1 arrives at about the time the water-rescue unit arrives and assumes Command.

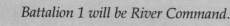

Battalion 1 will be River Command.

Battalion 1 confers quickly with the dive master. It is determined that the fire boat will pick up the two divers (the rescue diver and the backup diver) and the dive master at the nearby dock and that they will work off the fire boat. Command now needs to focus on the rest of the incident. The dive master is in charge of the dive.

On shore, Command has the safety officer set up a Warm Zone on the shoreline to keep onlookers back from the water's edge. Command asks that a police crew be sent to the King Bridge to watch the civilians who have gathered there. (Focus!)

Medic 2 has asked Command to get three private ambulances to transport the three men who are out of the water. Seven minutes after leaving shore, a single diver enters the water to begin a search. Anchor lines are in place, the rescue diver is tethered, and the backup diver is focused on the rescue diver's bubbles. The dive master looks over the area to ensure safety. Several boats are drifting and idling in the area.

Dive master to Command: Can you get the police boat to clear the area around the dive site for our safety?

Command: Okay.

Command to Dispatch: Contact the police boat, and have them maintain 100 yards in all directions around the dive area.

After approximately six minutes underwater, the diver signals that he has something and heads to the surface. The last victim is pulled into the fire boat and taken to shore, where Medic 2 is waiting to transport the victim.

Medic 1 checks the diver and the backup diver, and the members assigned to dive support start to pick up equipment.

While this is being done, Command briefs the media personnel who have gathered. Soon, all units are picked up and sent back for rehab.

More than one dive master may be needed at incidents involving several victims. In that case, an Operations Section would have to be established. The individual dive masters would report to Operations.

Regardless of the size or complexity of the dive or the type or condition of

Figure 23-4. Activity at a marina during a dive operation. Medics in the background are working on the victim. Police are questioning witnesses. (Photo by Ken Anello.)

the water at the time of the dive, the incident management system is critical to a coordinated, controlled incident. It brings control and focus to a very stressful and chaotic event. Everyone has a role and responsibility—a specific place to channel their energy.

BOMB INCIDENTS

Dispatching fire units to a bombing may not be an everyday occurrence, but it could happen. The bombing of the World Trade Center in New York City and the bombing of the Alfred P. Murrah Building in Oklahoma City, Oklahoma, have opened the fire service's eyes to this ever-present threat.

One of the stumbling blocks in preparing for a bombing incident is determining the line of authority. Police and fire units will respond to incidents involving bombs—whether a bomb has exploded or not. The questions to be answered are, "Who will take the lead at an actual incident? Who will be in charge?" Unified command may help resolve this issue. I believe that time will provide the answers after careful examination of the above two incidents have been completed. The fire department took command at both of these incidents, and both incidents had problems because unified command was not used. I believe that unified command is getting a closer look as a result of these two incidents.

Authority

In my opinion, determining who has authority at a bomb incident is quite simple. If the bomb has not exploded, the police should be in charge; they then would be the lead agency at the incident. (Notice that I refrained from saying that the police would be Command. To date, only a handful of police departments use any form of incident management.)

If the bomb has exploded, the incident must be treated as a mass-casualty incident (with or without an ensuing fire). As such, the fire department should be in charge until all victims have been accounted for and any associated fire, haz-mat, or other rescue concerns have been addressed. Then, the situation could be turned over to the police.

Unexploded Bomb Incidents

When responding to a report of a bomb that has not exploded, procedure should dictate the level of response. Most departments send a full (regular) alarm assignment of two or three engines, one truck (and possibly a heavy rescue squad), and a battalion chief. I do not know the level of response sent by the police, but I do know that it is significant.

On arrival, fire units should stage at least one block from the involved structure and await for direction from the police. On-scene crews should be prohibited from using radios. They would be able to receive messages but not transmit anything over the radio (some devices are radio-frequency sensitive). The chief may walk up to the police command post (if there is one) and inform the officer in charge that he is on the scene and awaiting direction. Policy should state that firefighters *do not look for explosive devices.* We are not trained or equipped to do this.

Our functions are to remove any victims should an explosion occur and to extinguish any fire that may ensue. That's all. We should not enter the structure or area prior to (1) the bomb's exploding or (2) the "All clear" has been given by the police, indicating that a search has been conducted and that no devices were found. The only reason fire personnel should enter a structure

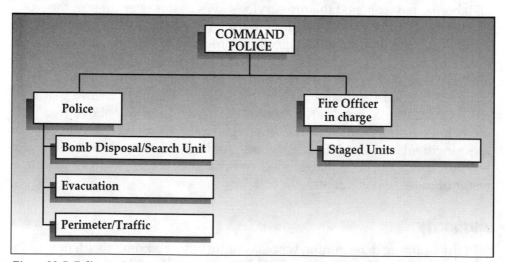

Figure 23-5. Police are in command of this incident, which involves an unexploded bomb.

or area before the police search has been completed is to assist in evacuation. If evacuation should be necessary, the search should be rapid and thorough, and personnel should leave the building quickly. After the evacuation has been completed, firefighters should return to their apparatus and await direction. (If fire personnel are to conduct an evacuation, sector assignments should be given in a face-to-face setting, not over the radio. Officers should be told to turn their radios off or to leave them outside at their rigs. Once their specific sector has been evacuated, they should report to a clearing area outside for reassignment or to be sent back to their rig.)

Exploded Bombs

The fire department should be in command when responding to a structure in which a bomb has exploded. Command's first concern should be the stability of the remaining structure. He must review his priorities on arrival. In addition to firefighter safety, which must always be paramount, Command will have to consider whether savable victims are inside the structure and the safety of the civilian rescuers who arrived at the scene before the fire units. For the most part, these incidents must be treated as mass-casualty incidents.

The areas of concern at such incidents include the following:

- *The weakened structure.* Police may not recognize this hazard. They may never have had to focus on this problem before, and it may not even occur to them that the blast may have weakened the remaining portion of the structure. Some firefighters, too, may underestimate the power of the blast. Secondary collapse and explosion may occur. Unstable floors, ceilings, walls, and stairs pose additional threats to safety. Sector officers must resist getting involved in hands-on operations and focus on their crews' safety and environment. Structural engineers may have to be called to the scene to assess the structure's stability, to help select work sites, and to prioritize areas.
- *Ensuing fires.* The fire attack tactics normally used may be ineffective or unsafe for extinguishing fires caused by a bomb blast. Inaccessible areas of the bomb site may have to be left to burn. Operations, in conjunction with Command and the sector officers, will make such determinations.
- *Rescues.* Generally, four types of victims are present at bombing incidents:
  —the uninjured who must be safely guided out of the unstable structure,
  —the slightly injured, who can move under their own power with a little assistance,
  —those who cannot move on their own but do not need extrication, and
  —those who must be extricated and who may or may not need assistance to leave the area.

  Priority must be given to victims who have the greatest need for medical attention.
- *Extrications.* Victims may need to be cut from structural wreckage. Extrication must be approached carefully. Cutting or moving anything could cause a ceiling or other part of the structure to come down. Pushing

debris away could cause the floor to collapse.

There are two types of extrications—manual and mechanical. In manual extrications, rescuers can free entrapped victims without having to use heavy tools. Simply opening a door or removing light building materials may be sufficient. Obviously, mechanical extrication requires the use of equipment such a pneumatic cutters, air bags, torches, and so on.

- *EMS concerns.* Triage and packaging can be performed within the areas where victims are awaiting rescue. Backboards and other immobilization equipment may have to be taken into the damaged zone, and treatment may have to be administered in the danger zones. Raising and lowering systems may have to be used to expedite victim removal. A transportation officer may be required to coordinate transport vehicles within the area.
- *Method of approach.* Rescuers would enter the damaged area of a bombing site in the same manner as they would a structure damaged by fire: Approach from the area of least involvement and move toward the area of greater involvement.

UNIFIED COMMAND

As indicated above, some incidents may fall under the jurisdiction of both the fire and police departments. Fire personnel would concern themselves with any fire(s) and rescuing the injured. The police would be involved with the aspects of rescue and crime, including the preserving of evidence. Unified command is best suited to such incidents. Under this type of command, needs can be identified from the perspectives of both entities and the solutions proposed would be acceptable to both. The flowchart in Figure 23-5 depicts an incident that involves a police search for an unexploded bomb. As indicated, the police are in command. That's the way it is, and that's the way it should be. The fire side indicates that a fire officer (probably a chief officer) is in charge of the fire section and has staged units directly under him. If the Police Command issues no direction to the Fire Section Officer, the fire units will remain staged until the incident has been terminated or Police Command indicates that fire crews are no longer needed.

SCENARIO

The incident is in a two-story office building that houses eight small retail and business occupancies. Initial reports indicate that as many as 25 victims may still be in the building. Fifteen injured occupants are outside when we arrive. A small fire is burning on Division 2 of the structure. The blast weakened the center of the structure, but all four exterior walls are partially standing. Most of the roof is missing or has collapsed.

The flowchart in Figure 23-6 depicts an incident at which a bomb has exploded.

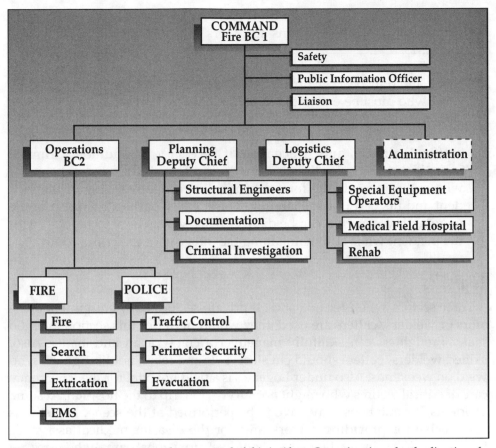

Figure 23-6. The fire department has command of this incident. Operations is under the direction of a fire chief officer, who has a police section under his direction. Note that the Planning Section also has "Police" in a Criminal Investigation Sector.

As indicated in Figure 23-6, the incident is under the command of the fire department. The operations section is headed by a battalion chief; a fire branch and a police section are under Operations. Fire branch has the following components:

- *Fire group.* They should be focusing on the fire on the second floor. Although not indicated, an attack group, a backup group, and perhaps a ventilation group could all be under the Fire Group Section Officer. In this situation, more than 25 companies could comprise the fire branch and not exceed the effective span of control (More companies could be used if any one of these sectors is also branched off.)
- *Search group.* This group would focus on locating viable victims. If only minor manual extrication is required, this group could help remove located victims. Once victims have been located, one or two fire crew members should be left with them to begin treatment. Depending on how many personnel are available and how the building is sectored off, more than 20 crews could be working under the search group.

- *Extrication group.* This group must focus on the manual and mechanical extrication of trapped victims and work closely with the search group and EMS.
- *EMS.* This group should focus on victim's needs, including triage, treatment, and transport. Victims must be triaged as they are located to determine who must be extricated first. The EMS section would also oversee packaging, which is part of treatment.
- *The police section.* This section (Figure 23-6) has three functional sectors: Traffic, Perimeter, and Evacuation, if necessary. In addition, Criminal Investigation is under the planning sector. This sector must exist somewhere. It represents the formal criminal investigative wing of the incident and fits best under Planning. The stability of the structure and the ongoing investigation can be discussed, as well as how to proceed with the incident while still maintaining the integrity of the crime scene.

Logistics

Three sectors are under Logistics. If private-sector heavy equipment operators or welders/cutters are used, they would fit under this section. Not too many firefighters can skillfully maneuver a crane or an end-loader. Also, skilled welders/cutters should cut steel cable and rebar, not firefighters who weld on weekends. Also under Logistics is a field hospital for the emergency care of critical victims who might not survive the trip to the hospital. At some incidents, amputations may have to be performed at the scene. Logistics is responsible for providing a work area for the disaster medical assistance team (D-MAT). Finally, Logistics should set up a rehab area where rescue workers can take a break, have fluids replenished, and have vitals taken. If this incident were to be under unified command, the flowchart might look similar to that in Figure 23-7.

In Figure 23-7, command is shared between police and fire representatives. There are two operations sectors—Police Operations and Fire Operations. Except for this difference, these flowcharts are similar.

If a structural engineer were represented under unified command, the flowchart would look like the one in Figure 23-8. One aspect I particularly like about adding an engineer to this scenario is that three entities are at the command post, providing a "tiebreaker," if you will, for resolving questions that might arise. Whereas a fire officer may lean toward personnel and civilian safety and a police officer may be concerned about preserving evidence, a more neutral individual, such as a structural engineer, would weigh additional considerations when making a decision. Whenever it is possible, it is desirable to have three unrelated entities jointly sharing responsibility under unified command. It may prove more efficient.

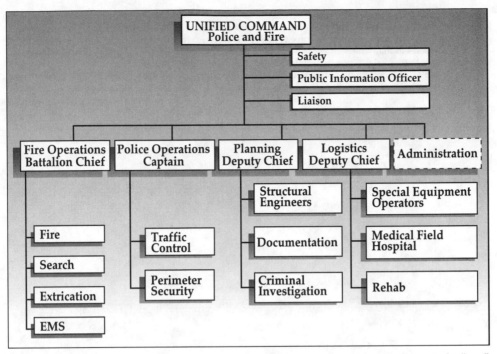

Figure 23-7. Unified command structure at a bomb incident. Note the relationship between the "two" Operations sectors.

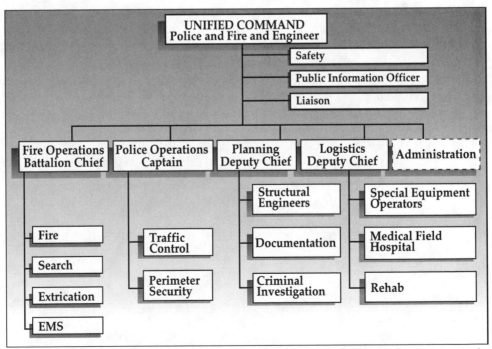

Figure 23-8. Unified command at a bomb incident. Command is shared by the fire department, the police department, and a structural engineer.

Endnotes

1. "Still box" is a term coined during the days when civilians called in fires in their neighborhoods through telegraph alarm boxes located on city streets. The boxes were established prior to the widespread use of telephones in many Eastern and Midwestern cities. A still-box alarm was one that was received by a means other than an alarm box, hence the name "still-" or "silent-" box alarm.

2. A dive master has training above that of the dive rescue team member. His basic task is to oversee the dive. He focuses on the safety of the divers and the search and recovery tactics they use.

REVIEW QUESTIONS

1. Why have fire department alarm system responses increased recently?

2. Which two of the four priorities of Command are of special concern at water-rescue incidents?

3. The command structure used at a water-rescue incident is similar to that used at which other type of incident?

4. What are the two types of bomb incidents?

5. Who should be in command at an incident that involves a bomb that has not exploded?

6. Who should be in command at an incident at which a bomb has exploded?

7. When fire units respond with the police on a call for the presence of a bomb inside a school, where should the fire units stage?

8. In an incident at which a bomb has exploded, where will the police investigation be incorporated into the command structure?

9. What are the two types of extrication at a bombing incident?

10. What type of command would be best suited to a major bomb blast in an occupied structure?

DISCUSSION QUESTIONS

1. Why does the author suggest that a separate medical sector be established for rescuers and injured civilians at several of the types of incidents discussed in this chapter?

2. How can communications be conducted at a suspected bomb site at which portable radios cannot be used?

3. Name two additional entities, other than a building engineer, that could be the third party represented at the command post in unified command.

Section V

Applying Incident Command

Applying Incident Command to Various Types of Incidents

The following case scenarios were designed to help demonstrate how what was discussed in the preceding chapters can be applied to various types of incidents. Assume that the Toledo (OH) Department of Fire and Rescue Operations is responding to these regular-alarm (or less) assignments.

Note that specific items and key phrases must be relayed over the radio to effectively communicate at the incident. Benchmarks are an important part of this equation. If assignments and other significant information are not given over the radio, they will not be documented. From a legal standpoint, a task that is not documented is a task that has not been performed.

I listen to other departments' operations often. The radio traffic patterns and terminology presented below are the same as those used in many other departments. I can pick out the progressive, safety-conscious departments in seconds. They have one thing in common—they use the incident management system at all incidents. What's more, they practice it and constantly try to improve on it.

If your department used incident command, you know what I'm talking about. It isn't something that you adopt overnight. You need to work with it and "massage" it into your department's operations. It's a tool. In fact, it's more like a toolbox, filled with little specific tools that can be pulled out when the situation requires. You may not always need a claw hammer or a Phillips screwdriver, but they're always there if you should need them. *The incident command "toolbox" is always with you, and you will always use some (or all) of the tools it holds.*

CASE 1: UNATTACHED RESIDENTIAL GARAGE FIRE

FIRE STRUCTURE

A residential garage in the south end of the city. The garage is not attached to the residence; it sits back 45 feet from the house at the end of a paved driveway. An alley runs directly behind the garage.

RESPONSE

Engines 5 and 9.

OTHER PERTINENT DETAILS

It is 0715 hours Satuday morning. The temperature is 36ºF (it is December). There is no precipitation. The ground and streets are dry.

ARRIVAL/ON-SCENE REPORT

At 0717, Engine 5 arrives. The officer gives an on-scene report and takes command.

Engine 5 at 157 Segur. We have a garage going good. Engine 5 is Segur Command.

Figure 24-1. Garage fire with heavy involvement in the rear.

SITUATION ON ARRIVAL

The officer of Engine 5 finds a 16- × 20-foot wood-frame garage well involved. The fire is in the rear and venting out a window and door.

COMMAND'S PRIORITIES

- *Firefighter safety.* Command believes his crew can make a safe interior attack on this fire. The span of communication is narrow, as is the span of control. Crews will not be very far from their egress.
- *Civilian safety.* At this time, it is not known whether a civilian is inside the garage. However, until the fire is knocked down, a search would be useless and unsafe due to the size of the fire in relation to the size of the structure.
- *Stop the problem.* In this case, extinguishing the fire will stop most of the other problems.
- *Conserve property.* Overhaul will be needed as soon as a search has been conducted.

INITIAL ASSIGNMENTS

Engine 5 makes the following assignments:

Dispatch: Engine 5 will be Attack. Engine 9, stage near the alley. Okay, Command at 0718.

INFORMATION GATHERED AT THE SCENE

The owner tells Command he started his car about 15 minutes ago to let it warm up. He went back into the house. He was inside about 10 minutes when he noticed the fire.

At this time, Command knows the following:

- The rear portion of the garage is fairly well involved in fire, near the alley (Side 3 of the garage).
- At this time, there are no exposure problems.
- According to the occupant, no one should be in the garage.
- The cause of the fire is unknown.
- He has started an interior attack on the garage from Side 1 with a 1¾-inch line. At this time he is working off the booster tank.
- He has an engine staged, if it should be needed.

SECONDARY ASSIGNMENTS

Command considers a backup sector. The area of the garage is small, and the attack crew can easily exit the garage. Nevertheless, Command has Engine 9 come up to the fire and pull a backup line off its engine (alternate water supply).

Command to Engine 9: Pull a backup line off your engine. Your assignment is backup.
Engine 9: Okay on backup.

The line is not charged. By this time, the front and rear doors of the garage are open and both attack crew members are visible. Engine 9 officer leaves one man on the line and the driver at the pump and reports to Command.

Command wants to ensure that the fire is being contained in the garage. He has the officer of Engine 9 check out Sector 3. Command gets his first report from Attack.

Figure 24-2. An initial attack at the garage fire. Engine 9 staged near the alley.

Attack to Command: We have the fire knocked down.

Command: Okay. Give me a search in there.

Attack: Okay.

Dispatch: Engine 5 is Search.

Command: Okay, Command at 0723.

Sector 3 to Command: There is no exposure problem in the rear of the garage. It looks like they got it from here. Do you want me to bring up my crew to assist in overhaul and salvage?

Command: Okay. Engine 9 is now Overhaul.

Dispatch: Okay on Engine 9 as Overhaul at 0725.

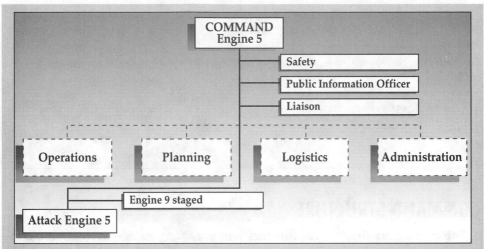

Figure 24-3. *Garage fire prior to the assignment of Engine 9.*

Command now must tell the dispatcher the status of the situation on Segur:

Command to Dispatch: We have the fire knocked down and under control. I won't need any more units here. I'll get you an "All clear" in a minute. We'll be here about a half-hour with Overhaul.

Dispatch: Okay at 0726 hours.

The smoke in the garage is dissipating very well. There is no need to get a truck crew with a fan on the scene to do what Nature is handling. Search gives Command an "All clear" in a face-to-face report.

Command to Dispatch: I have an "All clear" on Segur.

Dispatch: Okay at 0727.

Crews work on overhaul and salvage while Command gets information from the owner.

Command has the officer of Engine 9 look for an area of origin. Command locates the owner and asks him about the cause of the fire. The occupant stated that he was warming up his car to go to the store. After he started the car, he came back into the house to let it warm up. A few minutes later, he noticed smoke coming from the garage. The officer of Engine 9 traces the burn patterns and the amount of damage back to the engine area of the car. Apparently the fire was caused by a mechanical failure of some kind. It was not a suspicious fire. Command reminds the occupant to call his agent if he has insurance, suggests that he make a list of any possessions that may have been damaged in the fire, and offers helpful information such as how to get the car towed if necessary and how to get a copy of the fire report for the insurance company.

At 0759, both engines are picked up, and the incident is terminated:

> *Engine 5 to Dispatch: We're leaving Segur Street. Segur Command is ter-minated.*

COMMAND STRUCTURE

The command structure at this incident was very basic—two engine companies operating under formal command. As indicated in Figure 24-4, the span of control was not exceeded at this incident. Command handled the staff positions as well as the functions of command. Command initially made one functional assignment (attack) to his crew. He also positioned Engine 9 and had it stage. Command is still responsible for the "ghost" sections and staff positions represented by dotted lines at this incident.

The previous flowchart indicates Engine 9's later assignment, checking Sector 3 of the garage. At a larger incident, subsequent assignments should be noted next to the companies that are working.

According to the time line for this incident, most of the appropriate sectors were assigned at this incident. Command was established. Attack and Overhaul were assigned. Somewhere between 0723 and 0725, Command reassigned Attack to Search. No time was given for this reassignment because dispatchers did not pick it up on the transmission. The "All clear" given by Command at 0727 would indicate that a search was done—but by whom? If incidents are not audiotaped or if the audiotapes are destroyed, it may be hard for the department to prove that a task had been done. If dispatchers are notified and transcripts are made of every working incident, much aggravation can be saved later.

The incident was terminated at 0759.

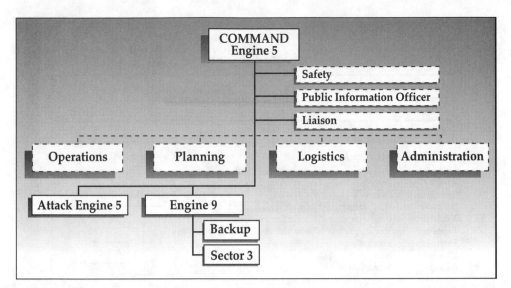

Figure 24-4. The same garage fire after the assignment of Engine 9.

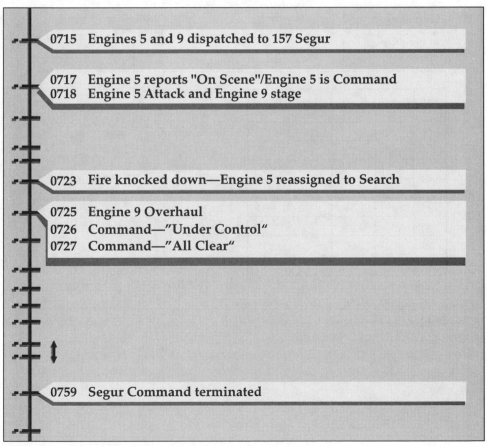

Figure 24-5. Incident time line.

SCENARIO #1

Incident Worksheet

Address: 157 SEGUR	Owner: Wm JENSEN	Occupant: N/A
	Address: SAME	Phone:
Detectors: Yes (No) Worked	Phone: 555-6743	
Cause: MECHANICAL / ENGINE AREA	Utilities Off—Gas Elec Water	

Functional Sectors:

Fire	Haz-Mat	Vehicular Accident
Attack /	Entry	Patient Care
Search /	Decon	Extrication
Backup /	H.M. Safety	Backup
Ventilation	Perimeter	Traffic Control
Exposures	Medical/Rehab	Transport
Extension	Evacuation	
Overhaul	Backup	
Salvage	Scene Safety	

Geographical Sectors:

Division=Floor
Base=2 floors below fire

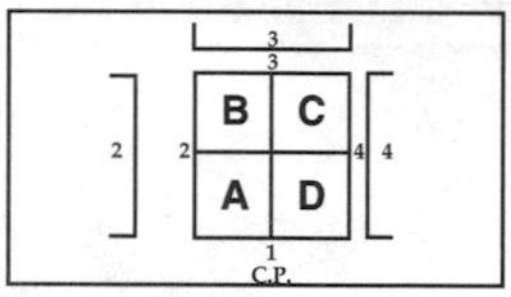

Fire Protection System
Sprinkler _____
Standpipe_____
Other _____
Connection Location
_____
Risers _____
Fire Pump Cap. _____
Misc. _____

Construction Type
Cons. Type W/F
Roof Type WOOD/TAR PAPER
Truss: _____
Occupancy GARAGE
Elevators_____
Misc. _____

Hazardous Materials
Material _____
I.D.# _____
L.E.L._____
U.E.L. _____
Specific Grav. _____
Vapor Dens._____
T.L.V. _____ ppm
Readings _____

Unit	Assgn# 1	Bench-mark	Assgn# 2	Bench-mark
E5	ATTACK	/	SEARCH	—
E9	BACKUP	/		
E9	SEC 3			

Sketch:

Figure 24-6. Incident worksheet.

CASE 2: TWO-CAR VEHICULAR ACCIDENT

INCIDENT DETAILS

Car #2, heading north on Route 206, ran a stop sign and was broadsided by Car #1, which was heading east on Route 101. The accident occurred on a Monday evening at 1934 hours.

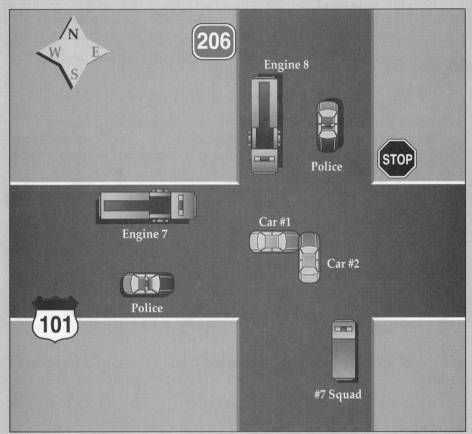

Figure 24-7. Initial placement of apparatus at this vehicular accident.

RESPONSE

Engines 8 and 7 are dispatched at 1935 hours. Engine 8 arrives at 1937 hours and reports, "Two cars involved at the intersection of Routes 101 and 206." Engine 8 takes command and assigns a quick triage to a crew member to determine the status of the victims.

COMMAND'S PRIORITIES

- *Firefighter safety.* The initial safety concern is traffic. After addressing that hazard, a backup line should be pulled before extrication. Members should be "gloved up."

- *Civilian safety.* Prior to opening any doors with mechanical extrication tools, a backup line should be pulled in case a flash fire should occur. Keep onlookers back.
- *Stop the problem.* Triage and treat those who can be manually extricated. With mechanical extrication, remove and treat trapped victims. Provide C-spine immobilization as soon as possible. Transport victims to health-care facilities.
- *Conserve property.* One car is not "totaled." Care should be taken not to damage it further. The other car is totaled and will not be a concern.

ASSIGNMENTS

Command assigns the remainder of Engine 8 to Car #2 and Engine 7 to Car #1.

Command to Engine 8: You're Patient Care in the car heading north. Engine 7, you're Patient Care in the eastbound car.

Dispatch: Okay on patient care, Engines 8 and 7 at 1938.

Command is told that the doors on Car #1 cannot be opened and that extrication is needed. Command special-calls for a heavy squad for extrication.

Route 206 Command to Dispatch: Send me a heavy squad for an extrication.

Dispatch: Okay on the heavy squad at 1940.

At this time, no hazardous fluids are on the ground, but since mechanical extrication is necessary, a backup line should be pulled. Only one victim, who is nearly packaged, is in Car #2. A line for backup can be pulled off Engine 8 and stretched for backup (Figure 24-8).

Command to Engine 8: Have your driver pull a backup line before the squad gets here.

Engine 8: Okay on a backup line.

Both sector officers inform Command that there are no critical patients and that three transport units will be needed. Knowing the location of the responding transport units and that there is a need for extrication, Command decides to have the ambulances stage on Route 101, east of the accident.

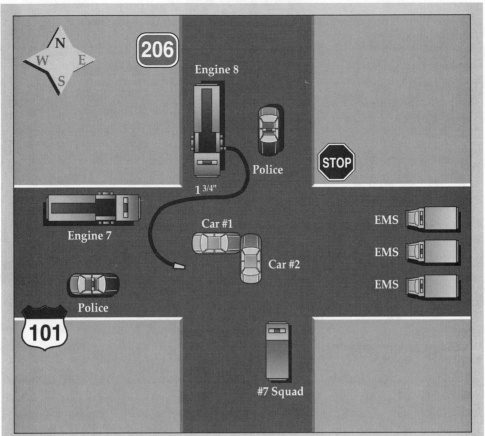

Figure 24-8. Several minutes into the incident, a backup line is pulled, and transport units are staged, awaiting direction to "come up."

Command to Dispatch: I need three ambulances. Have them stage on Route 101, east of the accident.

Dispatch: Okay on three ambulances and staging on 101 east of the accident at 1943.

All victims are extricated at 1958.

Command to Dispatch: All victims have been extricated and are being packaged.

The last victim is placed in the ambulance at 2012.

Command to Dispatch: All victims are transported. These units are picking up and will be in service shortly. Route 206 command is passed to the State Police.

Dispatch: Okay at 2012.

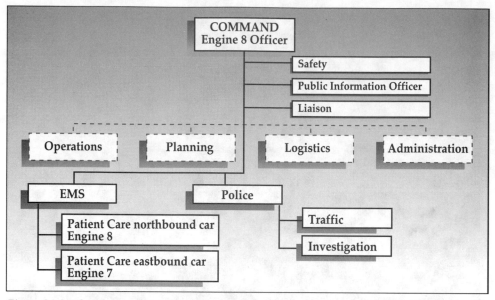

Figure 24-9. The command structure above depicts initial assignments at this accident. Note that "Police" is under the command of the incident commander.

COMMAND STRUCTURE

Initially, Engine 8 takes command and assigns patient care to both cars. As indicated in Figure 24-9, two sectors are directly under Command—EMS and Police. There may be concern about placing the police under the command of a fire officer; incorporating the police in a flowchart is not a normal occurrence. But think about it: The fire officer on the scene is Command and in charge (by state law in some states). If a police officer is injured, can Command be held liable? If Command observes that a police officer is doing something counterproductive to the incident and does not take steps to stop or change it, can he be held responsible? Is there a guarantee that he won't? If the police are on the scene working with the fire personnel, I believe that they must be represented somewhere on the command chart. The police are handling the traffic and the investigative phase of the incident.

INCIDENT TIME LINE

According to the incident time line, the first fire unit arrived on the scene at 1937—the time the fire department actually began its role in the incident. Some may argue that fire department involvement begins on notification. That is true. However, I believe that if we are notified of an incident (either real or imagined) and respond and arrive within acceptable time frames (as we did in this instance), that information may be insignificant. As indicated by transcripts of the incident:
- Command was established.
- Patient care was assigned to two cars.

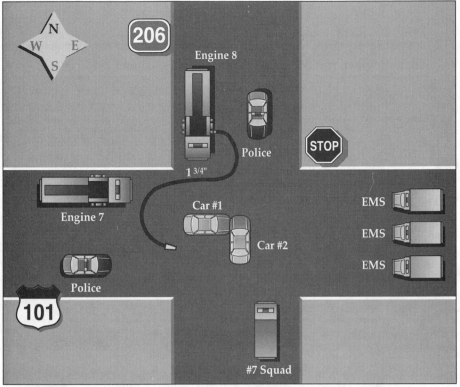

Figure 24-10. Final assignments at the vehicular accident. Backup lines are pulled and transport vehicles staged.

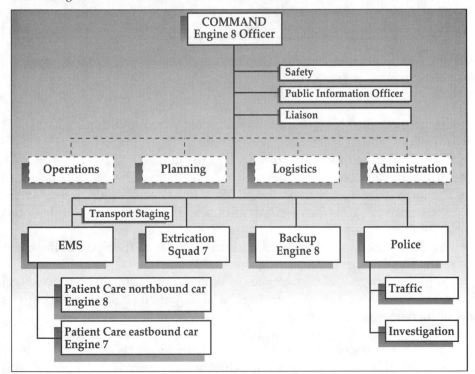

Figure 24-11. Final command structure of the vehicular accident.

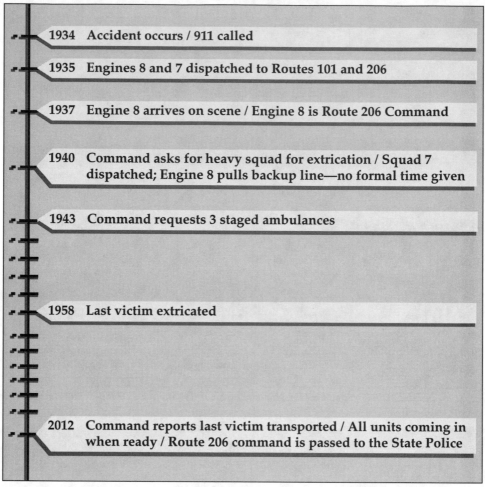

1934   Accident occurs / 911 called

1935   Engines 8 and 7 dispatched to Routes 101 and 206

1937   Engine 8 arrives on scene / Engine 8 is Route 206 Command

1940   Command asks for heavy squad for extrication / Squad 7
       dispatched; Engine 8 pulls backup line—no formal time given

1943   Command requests 3 staged ambulances

1958   Last victim extricated

2012   Command reports last victim transported / All units coming in
       when ready / Route 206 command is passed to the State Police

Figure 24-12. Incident time line.

- Command called for extrication equipment when told that a victim was trapped.
- Command called for transportation for the injured.
- All victims were extricated.
- The last victim was transported. (This places control of all victims in the care of private ambulance companies, and time frames can and should be established for this.)
- Fire command was terminated and turned over to the State Police prior to the fire department's leaving the scene.

Most of the reports and occurrences indicated on the time line are based on direct conversation between Command and dispatchers. Some notifications and occurrences were the result of the dispatchers' listening in and picking up on subsequent assignments, which is acceptable if the dispatchers are not too busy. There may be times when dispatchers are too busy to listen attentively to all radio transmissions not directed specifically to them. If you

Incident Worksheet

Address: *Rt 206*	Owner: *CAR#1* Address: *VICTIM#1* Phone: *CAROL BADER* *F-40 DOB 2/11/56* *AGE VERMONT*	Occupant: *CAR#2* Phone: *STEVE JONES* *M-24 · 177 MAIN ST.* *WATERVILLE*
Detectors: Yes No Worked		*DOB 9/22/62* *McGANDER HOSP.*
Cause:	Utilities Off—Gas Elec Water	

VICTIM 2 ANN BADER
F4 - DOB 1/7/92 SAME ADD

Functional Sectors:

Fire	*Haz-Mat*	*Vehicular Accident*
Attack	Entry	Patient Care
Search	Decon	Extrication
Backup	H.M. Safety	Backup
Ventilation	Perimeter	Traffic Control
Exposures	Medical/Rehab	Transport
Extension	Evacuation	
Overhaul	Backup	
Salvage	Scene Safety	

Geographical Sectors:

Division=Floor
Base=2 floors Below fire

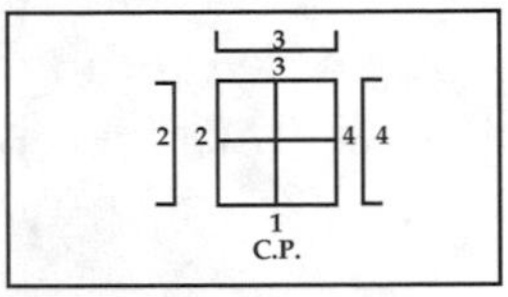

Fire Protection System
Sprinkler _____
Standpipe _____
Other _____
Connection Location

Risers _____
Fire Pump Cap. _____
Misc. _____

Construction Type
Cons. Type _____
Roof Type _____
Truss: _____
Occupancy _____
Elevators _____
Misc. _____

Hazardous Materials
Material _____
I.D.# _____
L.E.L _____
U.E.L. _____
Specific Grav. _____
Vapor Dens. _____
T.L.V. _____ ppm
Readings _____

Sketch:

Unit	Assgn# 1	Bench- mark	Assgn# 2	Bench- mark
E8	*CAR 2* *PAT CARE*	✓	*BACKUP*	
E7	*CAR 1* *PAT. CARE*	✓		
Sq7	*EXTRICATION*	✓		
MEDCORE	*VIC 1*		*CAR1*	*RIVERSIDE HOSP*
MEDCORE	*VIC 2*		*CAR1*	*" "*
WALKER	*CAR2*			*McGANDER HOSP*

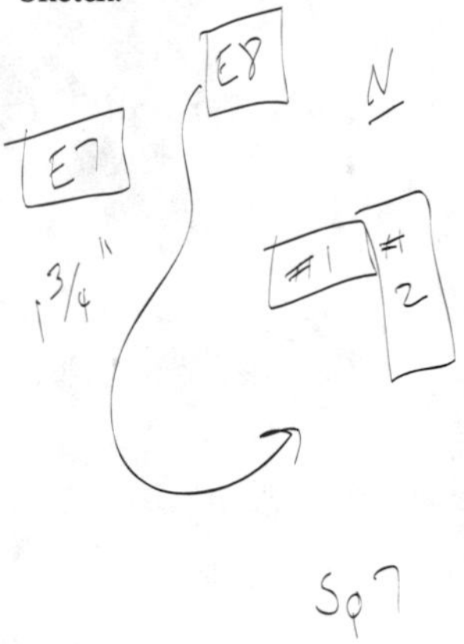

Figure 24-13. Incident worksheet.

believe it is important that an entry be in the time line, then direct a transmission to dispatchers. That way, you will be sure to receive a time reference for important incident occurrences.

CASE 3: SINGLE-FAMILY RESIDENTIAL FIRE

FIRE STRUCTURE

The structure is a single-story, one-family frame residence (platform) in a newer subdivision of the city. The home is occupied by four family members. The fire started in the front bedroom in Sector A.

ADDITIONAL DETAILS

The mother and the two preschool children were at home at the time of the fire. The mother heard smoke detectors going off at about the same time the two children came running into the kitchen. The mother and the two children fled through the back door and went next door to call 911. The fire department received the call for a residential structure fire at 1547 hours.

RESPONSE

Engines 5 and 6, Truck 5, Squad 4, and Battalion 1 are dispatched. All units are en route at 1548.

SITUATION ON ARRIVAL

Engine 5 arrives on the scene and observes heavy smoke and fire coming from the front of the house. Engine 5 wrapped a hydrant while pulling up to the house but did not leave a man at the hydrant.

Engine 5 is at 1516 Page. A one-story frame is well-involved. The next unit in westbound on Page, charge the hydrant for us. Engine 5 is Page Command.

Dispatch: Okay. Engine 5 reports a working fire. Engine 5 is Page Command at 1550.

Table 24-1. RESPONDING UNITS, PERSONNEL ASSIGNMENTS, RESPONSE TIME

Unit Dispatched	Members Assigned	Approximate Response Time
Engine 5	Four	60 seconds
Engine 6	Four	2 minutes, 30 seconds
Truck 5	Four	1 minute, 15 seconds
Squad 4	Four	1 minute, 30 seconds
Battalion 1	One	3 to 4 minutes

The mother runs to the officer and tells him that everyone is out. Engine 5 considered going "fast attack" initially; however, when he considered that the next-in units were only a minute behind him, he decided to take formal command.

COMMAND'S PRIORITIES

- *Firefighter safety.* Command believes his crew can make a safe interior attack on this fire. The structure is still "sound." With turnouts, there should be no problem.
- *Civilian safety.* Reports from the occupant indicate that everyone is outside. A search is needed, but its place on the to-do list can be a bit more flexible and thus can be lower.
- *Stop the problem.* The fire is confined to the front portion of the house. Heat and smoke are spreading up and toward the rear of the structure. A hoseline must be positioned to stop the spread of fire and then to extinguish it.
- *Conserve property.* Immediate ventilation is needed to channel heat and smoke up and out of the house. Salvage and, finally, overhaul are needed. At this time, the amount and location cannot be judged.

COMMAND'S TO-DO LIST

Command quickly looks at the picture in front of him and reviews his priorities. Due to the factors involved, the units dispatched, their response time, and the conditions observable on arrival, Command establishes the following to-do list:

- Confine the fire to the area of involvement by a properly placed *attack* line.
- *Ventilate* to remove the products of combustion from the structure.
- Protect interior crews by placing a *backup* line in the appropriate area.
- *Search* the structure for any possible victims.
- Check for *extension*.
- *Salvage* as needed.
- *Overhaul* as needed.

ASSIGNMENTS

According to Command's to-do list, the following assignments are given to units already on the scene:

Command: Engine 5 is Attack. Truck 5, you're Topside Ventilation.

Dispatch: Okay on attack and ventilation at 1551 hours.

At this time, no more units are on the scene. As stated in Chapter 8, assigning sectors (the "Big Four" or others) to dispatched units that are not yet on the scene may prove to be counterproductive. The units may become involved in an accident and never reach the scene or be delayed by traffic or a passing train. Another unit may arrive before the one Command expected to be on-scene first. It's best that Command wait until a unit arrives on the scene before he gives an assignment so that he can be sure that the next task assigned will be the top one on his to-do list.

As Command hears the squad pull up, he quickly reevaluates his to-do list based on the following:

- current conditions,
- previously assigned sectors, and
- anticipated incident requirements based on his current to-do list.

Command's new to-do list looks something like this (note that the two previous items on the list have been removed and the items lower on the list have been moved up toward the top):

- Protect interior crews by placing a *backup* line in the appropriate area.
- *Search* the structure for any possible victims.
- Check for *extension*.
- *Salvage* as needed.
- *Overhaul* as needed.

The next item on the list is to protect interior crews. The last "life" or safety concern is to search. We still have an accounting of all occupants. At this time, it is more important to assign a backup sector than to begin a search.

Command to Squad 4: You're Backup.

Squad 4: Okay on backup.

Command now knows the following:

- The fire is in the front section of a one-story frame home.
- All occupants are apparently out at this time.
- A two-person crew (Engine 5—the officer is Command, and the driver is at the pumps) is attacking the fire.
- A truck crew is opening up the roof.
- A crew (Squad 4) is providing a backup line for the interior crews.

Command hears Engine 6 pull up and report on the scene. At this time, Command again has to quickly review his to-do list. No big changes. His next item is still "search."

Command to Engine 6: You're Search.

Engine 6: Okay on search.

About this time, Attack gives his benchmark:

Attack to Command: We have the fire knocked down.

Command: Okay. I'm changing your designation to Extension.

Engine 5: Okay on extension.

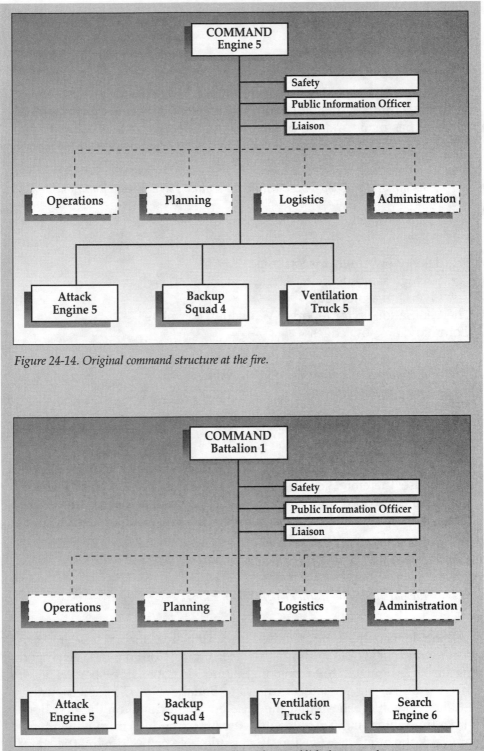

Figure 24-14. Original command structure at the fire.

Figure 24-15. Command structure after Battalion 1 has established command.

Command to Dispatch: Fire knocked down. Engine 5 is now Extension.

Dispatch: Okay at 1555.

Battalion 1 arrives and assumes command. He gets a face-to-face briefing from Command (the officer of Engine 5), who reports on the assignments he has made and the progress of the fire. Battalion 1 wants to get the officer of Engine 5 back to his crew, who are now inside as Extension.

Command to Dispatch: Battalion 1 is now Page Street Command.

Dispatch: Okay. Battalion 1 is now Command at 1556 hours.

A quick review of what is left to do:
- Salvage as needed.
- Overhaul as needed.

The Ventilation sector reports "Ventilation complete."
Command acknowledges:

Command: Okay on ventilation complete. Truck 5, I'm changing your designation to Salvage.

Truck 5: Okay on salvage.

Search reports an "All clear." Command doesn't want too many people in the house; therefore, he tells Search to come out and start to pick up.

Command also looks at the picture and observes that the majority of the smoke is dissipating. In Command's mind, the fire is under control.

Command to Dispatch: This fire is under control.

Dispatch: Okay on under control at 1558.

All incidents of this type are based on what a "reasonable, prudent" fire officer would or should be doing. In this instance, Command has changed Engine 5's assignment to Extension. Engine 5 is now a three-person crew, since the officer has passed command to Battalion 1 and joined his crew. Battalion 1 would believe that the hoseline is left down on the first floor with a firefighter on the nozzle while the officer and his third crew member check the attic for extension. If a scuttle hole or pulldown attic stairs are used for access to the attic, the crew member on the nozzle will ensure the safety of the remainder of the crew, who actually check the attic. If heavy smoke conditions are present in the attic or if open flame is visible, it would be assumed that the officer assigned as Extension would inform Command and be reassigned as Attack.

Now there is little need for a backup sector. Engine 5 finds no extension in the attic. Command reassigns Backup to overhaul in the fire room.

> *Command to Backup: I'm changing your designation to Overhaul. Trace the spread back to the room of origin, but hold off overhauling in the area of origin until I tell you.*
>
> *Squad 4: Okay on overhaul.*
>
> *Command to Dispatch: Squad 4 is now Overhaul.*
>
> *Dispatch: Okay at 1600 hours.*

Command goes mobile and finds the occupant. He begins to discuss the fire and how it started. It is believed that the fire was started by the children, who were playing with a cigarette lighter. Command calls for an investigator to determine the actual cause. Search gives Command an "All clear."

> *Command to Dispatch: We have an "All clear." Also, send me an investigator on Page Street.*
>
> *Dispatch: Okay at 1601.*
>
> *Overhaul to Command: The area of origin is near the bed in the bedroom in Sector A. Looks like the kids' room.*
>
> *Command: Okay. I've got an investigator coming, so hold up with overhaul in that room.*
>
> *Overhaul: Okay.*

Again, Command reviews his to-do list. It seems as if all the needs of the incident are being met. The checklist on the command worksheet is reviewed. The sheet reminds him that he needs to call the electric company to cut the drop to the house; otherwise, the fire department could be held liable if the occupant should receive a "poke" or if wiring damaged by the fire should cause a short.

> *Command to Dispatch: We need the electric company out here.*
>
> *Dispatch: Okay at 1610.*

The fire is out. There was no fire spread outside the original fire room. The ventilation hole is covered with tar paper. Salvage covers are spread. Debris is shoveled out. The investigator takes photos of the area, talks to the family, and determines that the fire was accidentally started by children

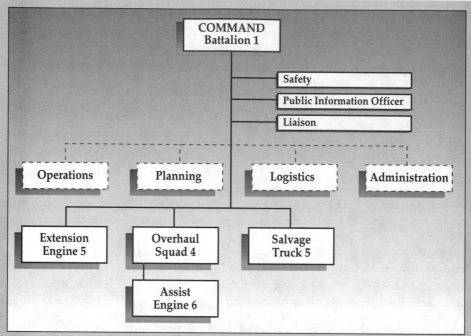

Figure 24-16. Command structure for final overhaul and salvage.

playing with a lighter in their bedroom. Engine 5 is picking up its lines, and the truck crew is taking a break. Battalion 1 gets ready to leave. He briefs the officer of Engine 5 and tells him about the electric company's coming, where to put the debris, and to make sure they take a walk through the house one last time to be sure the fire is out before they leave.

> *Battalion 1 to Dispatch: I'll be coming in on Channel 1. Engine 5 will be Page Street Command.*
>
> *Dispatch to Battalion 1: Okay at 1632.*

Command (Engine 5) now has Engine 6, the truck, and its crew to pick up. Approximately 35 minutes later, they take one last walk through the house and close the incident.

> *Dispatch: All units will be coming in on Channel 1. Page Street Command is terminated.*
>
> *Dispatch: Okay. Page Command terminated at 1705.*

COMMAND STRUCTURE

As Figure 24-14 indicates, the Engine 5 officer originally established command at this fire. Initially, Command made three functional sector assign-

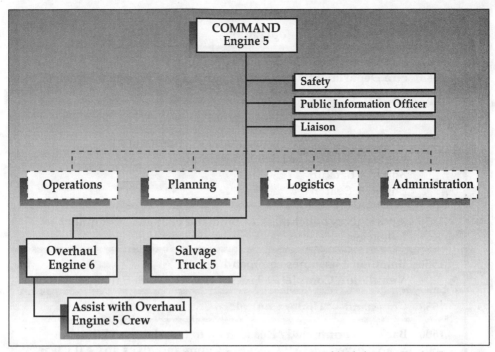

Figure 24-17. Command structure after Battalion 1 passes command back down to Engine 5.

ments at this fire. The span of control was not a concern during the initial assignments of this incident. Battalion 1 arrived and took Command from the officer of Engine 5. By the time Battalion 1 established command, all of the "Big Four" had been assigned (see Figure 24-15). As the fire was brought under control, Command began to reassign the on-scene units to the second phase of this incident. As seen in Figure 24-16, Extension, Overhaul, and Salvage are now the working sectors at the incident. When Battalion 1 left at 1632, Command was given back to the Engine 5 officer, who had Engine 6 and the remainder of Command's crew on overhaul. Some activities not associated with the structure were going on at the time, including picking up hoselines in the street and cleaning tools at the scene. These acitvities need not show up as assignments or sectors. They basically have no relevancy to any part of the incident except internally. (Internal SOPs should specify procedure for such activities.) Truck 5 is finishing salvage. Figure 24-17 represents the final command structure of the incident.

INCIDENT TIME LINE

This incident looks relatively good under the incident time line. There was little lag or time lost between dispatch and arrival time. Sector assignments flowed in a timely and justifiable manner. There was a smooth transition between incident commanders. All of the "Big Four" functions were assigned, as were the second-phase assignments of salvage and overhaul.

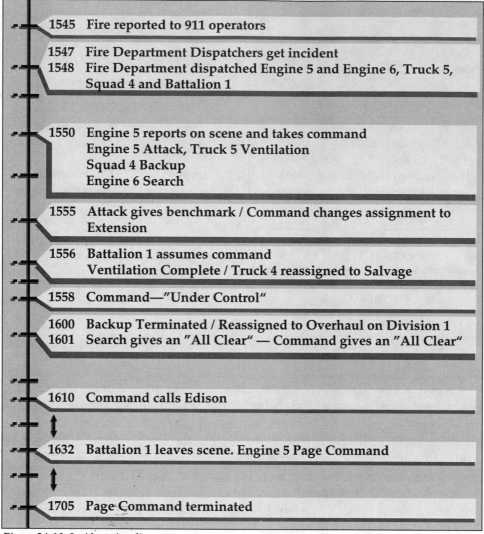

1545	Fire reported to 911 operators	
1547	Fire Department Dispatchers get incident	
1548	Fire Department dispatched Engine 5 and Engine 6, Truck 5, Squad 4 and Battalion 1	
1550	Engine 5 reports on scene and takes command Engine 5 Attack, Truck 5 Ventilation Squad 4 Backup Engine 6 Search	
1555	Attack gives benchmark / Command changes assignment to Extension	
1556	Battalion 1 assumes command Ventilation Complete / Truck 4 reassigned to Salvage	
1558	Command—"Under Control"	
1600	Backup Terminated / Reassigned to Overhaul on Division 1	
1601	Search gives an "All Clear" — Command gives an "All Clear"	
1610	Command calls Edison	
1632	Battalion 1 leaves scene. Engine 5 Page Command	
1705	Page Command terminated	

Figure 24-18. Incident time line.

However, Command failed to get specific time checks for some of his assignments. If a transcript is made of this incident, it can be determined that these sector assignments were made, but time frames would have to be estimated. If an attorney were to review this time line, he would find that all the bases necessary for an incident of this scope had been covered but that the time frames were incomplete.

Incident Worksheet

Address: *516 PAGE*	Owner: *John W. Smith* Address: *SAME* Phone: *555-1279*	Occupant: *SAME* Phone: *SARAH F34* *KIMBERLY F7* *JENNIFER F4*
Detectors: Yes No (Worked)		
Cause: *CHILD W/LIGHTER* *KIMBERLY*	Utilities Off—(Gas)(Elec) Water	

Functional Sectors:

Fire	*Haz-Mat*	*Vehicular Accident*
Attack ✓	Entry	Patient Care
Search ✓	Decon	Extrication
Backup ✓	H.M. Safety	Backup
Ventilation ✓	Perimeter	Traffic Control
Exposures	Medical/Rehab	Transport
Extension ✓	Evacuation	
Overhaul	Backup	
Salvage ✓	Scene Safety	

Geographical Sectors:

Division=Floor
Base=2 floors below fire

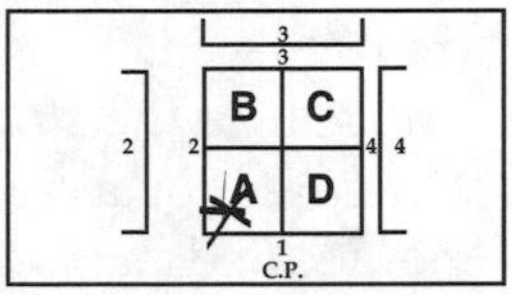

```
        3
        3
   2  2  B   C   4  4
        A   D
        1
       C.P.
```

Fire Protection System

Sprinkler _____
Standpipe _____
Other _____
Connection Location
_____
Risers _____
Fire Pump Cap. _____
Misc. _____

Construction Type

Cons. Type *W/F*
Roof Type *W/TAR. SHINGLE*
Truss:
Occupancy *411 SINGLE*
Elevators
Misc. *EDISON CAT*
DROP-UNTI
177

Hazardous Materials

Material _____
I.D.# _____
L.E.L. _____
U.E.L. _____
Specific Grav. _____
Vapor Dens. _____
T.L.V. _____ ppm
Readings _____

Unit	Assgn# 1	Bench-mark	Assgn# 2	Bench-mark
E5	ATTACK	-	EXTENSION	-
T5	VENT	-	SALVAGE	-
SQ4	BACKUP	-	OVERHAUL	✓
E6	SEARCH	✓	✓	
ALL CLEAR		UNDER CONTROL	-	

Sketch:

Figure 24-19. Incident worksheet.

CASE 4: FIRE IN A BALLOON-FRAME STOREFRONT

FIRE STRUCTURE

At 2146 hours on a Saturday in July, a report of a smoke odor at 1905 Vermont Avenue comes in to fire dispatchers from 911 operators. The structure, in an older neighborhood of the city, is an old two-story, balloon-frame storefront occupancy built at the turn of the century. It has a flat roof over a cockloft. The lower level of the building is occupied by a convenience store. Two aisles of shelving stocked with food and other products are in the front; a large storage area and cooler are in the rear of the building. The building has a full basement. Four apartments are above the store. Three are occupied. The only access to the apartments is a wooden stairway in the rear of the structure. An alley runs behind the building.

The structure is on a corner lot and has one exposure—an occupied nursing home, located next door to the storefront. The nursing home, an old converted house, has 16 rooms that had been converted to sleeping rooms, a gathering hall, and a kitchen/dining room area. At the present time, 14 occupants and a weekend staff of three are within the facility.

The temperature is 93°F; the humidity is very high.

RESPONSE

Fire dispatchers send Engine 7 at 2146 hours. The crew is familiar with 1905 Vermont. All are veteran firefighters and have been assigned to the #7 station for a long time. They have made frequent calls to this address, mostly for EMS runs for apartment occupants and store patrons.

INFORMATION GATHERED ON-SCENE

Workers in the store detected a slight odor of smoke coming from the storage area. They walked back there. They didn't see any smoke but could smell something burning. At about this time, they heard a customer in the store and returned to the front area. Several minutes later, they again smelled smoke and decided to call the fire department. After they made the call, one of the employees went back into the storage area to see if he could find anything. The other employee remained in the front to help customers. Within a short time, the employee who went to look for the source of the fire ran back up to the front of the store to report that fire was in the walls of the storage area. Everyone left the store through the front door at about the same time that Engine 7 arrived.

ON-SCENE REPORT

Engine 7 arrives on the scene at 2149 hours and reports that light smoke is showing from the store.

Engine 7 is at 1905 Vermont. Light smoke is showing.

(Engine 7 is the only unit dispatched and is operating under informal command. As such, he need not establish formal command.) The employees tell the Engine 7 officer what has happened.

Engine 7 to Dispatch: This is a working fire in an occupied structure. Make this a regular (or first) alarm. Engine 7 is Vermont Avenue Command.

Dispatch: Okay on the regular alarm at 2150 hours.

Command now knows the following:
- He is the only unit at this incident.
- There is a fire (the magnitude of which cannot be judged at this time) in an occupied storefront with apartments above. Life safety is a high priority at this incident. At this time, the store is unoccupied. Command has no idea of how many occupants are above the fire.
- Help (the first alarm assignment) is at least one to two minutes away.
- His crew consists of four persons.
- All he has is a booster-tank water supply.
- He is Command and has a three-person engine crew left until help arrives.

COMMAND'S PRIORITIES
- *Firefighter safety.* Personnel are a concern. He has two firefighters, a driver, and himself. There is no doubt that the officer should establish formal command. The fire has entered the structure of a balloon-frame building. Its travel is unpredictable. At the time of arrival, firefighters could enter and operate only for a short time if fire suppression and backup efforts are not mounted immediately.
- *Civilian safety.* Three occupied apartments are above the storefront. The number of occupants who may be upstairs cannot be estimated at this time. Additionally, only one stairway—which is accessible— serves the second floor. Civilian safety is a major concern.
- *Stop the problem.* Remove the civilians and then stop the spread of fire. These are the two major initial problems.
- *Conserve property.* Overhaul will be a major concern after life safety has been accounted for. The spread of fire is unpredictable, and all areas of the building could be affected by heat, smoke, and fire. Salvage operations must be considered.

Command now quickly reviews his priorities. Dispatch informs Command that the remainder of his regular alarm consists of Engines 16 and 17, Truck 17, Squad 7, and Battalion 3.

Looking at the picture and reviewing what resources he has on scene and en route, Command believes he has two options:

1. Quickly take a line inside and try to find the fire and hold it until help arrives (this action will do nothing to remove occupants from the structure). Additionally, the fire is in the walls. The structure is balloon frame. Who knows where the fire is now and where it's going to end up?)[1]

2. Have two of his crew members don SCBA and quickly search and remove any occupants on the second floor. This action may allow the fire to get a stronger hold on the structure.

ASSIGNMENTS

Command tells his crew to mask up and to try to get everyone out of the second floor.

Command to Engine 7: You're Division 2 Search (life safety over stop the problem). *Act quickly. Don't let the fire cut you off. The stairs are the only way in or out.*

Command to Dispatch: Engine 7 is Division 2 Search.

Dispatch: Okay at 2151.

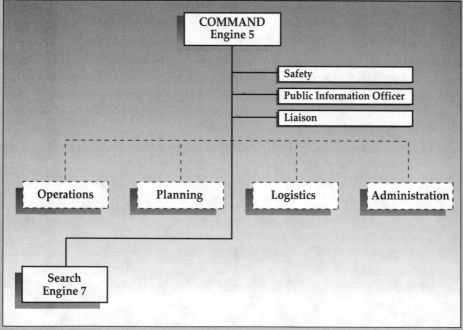

Figure 24-20. Command structure after Engine 7 officer asks for a regular alarm.

As Search rounds the corner toward the rear of the structure, Command hears Squad 7 pull up.

Command to Squad 7: You're Backup. Get a line on the rear stairway. I have two men on search on the second floor without a line. Be advised that you're working off the booster tank.

This is a very clear assignment. It paints a picture for the officer of Squad 7 and tells him exactly what his focus should be at the present time.

Command now knows that Backup is aware that water is at a premium, but a good crew can hold back a lot of fire with 500 gallons of water.

Dispatch: Squad 7 is Backup at 2152.

Command to Search: Be advised that Squad 7 is pulling a backup line on the rear stairs.

This should tell the search crew that once they hear the backup sector's benchmark, they can extend their search a little farther now that someone is watching out for them. Command gets a report from Search:

Search to Command: Tommy is bringing out three people from one of the apartments. Send someone up with me when you can.

Backup to Command: I'll send two of my crew up if it's okay. (This will leave the officer and one crew member on the backup line.)

Command: Okay.

Command to Dispatch: Send Vermont Command a medic unit.

Dispatch: Okay on a medic unit at 2154.

Backup to Command: Backup line is in place. Be advised that the fire is starting to roll in Sector B of Division 2 and it's in the cockloft now (keep Command informed).

As this transmission comes over the radio, Command sees Engine 17 and Truck 17 pull up to the corner.

Command to Engine 17: Lay into Engine 7, and then take lines off Engine 7. You're Division 2 Attack. Take your line up the rear stairs. The fire is in Sector B, and it's over your head in the cockloft.

Engine 17: Okay on Division 2 Attack, line in the rear.

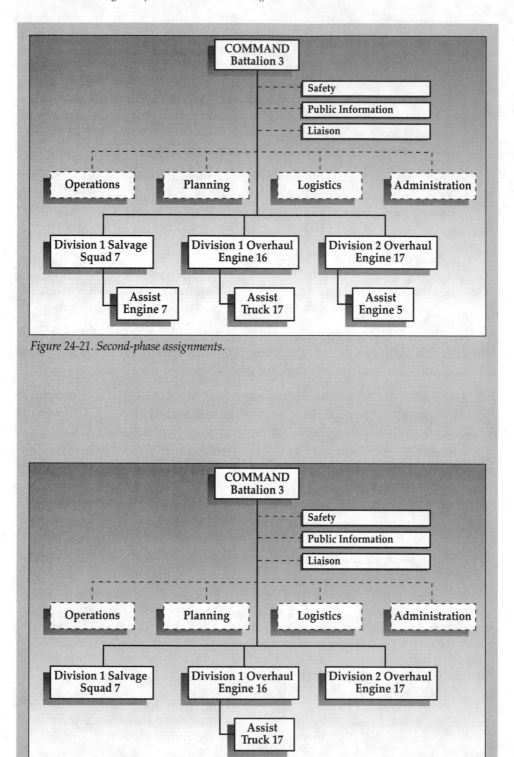

Figure 24-21. Second-phase assignments.

Figure 24-22. The command structure after Battalion 3 leaves the scene.

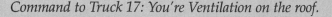

Command to Truck 17: You're Ventilation on the roof.

Truck 17: Okay.

Dispatch: Okay on Division 2 Attack and Ventilation at 2155.

Battalion 3 pulls up and gets a quick briefing from Command. Battalion 3 brings his clipboard on which to write the assignments.

Command explains: "I've got my crew on Division 2 as Search. Engine 17 is taking a line off us and is Division 2 Attack. Squad 7 is Backup in the rear stairs. The truck's going up to the roof to vent."

Battalion 1 takes Command.

Battalion 1: Battalion 1 is Vermont Command.

Dispatch: Okay. Battalion 1 Command at 2157.

Engine 16 is staged at Locust and Vermont.

Command to Engine 16: Lay in. You're Division 1 Attack. Take your line through the front.

Command knows that he still has things to do but has no companies left.

Command to Dispatch: Give me another engine and squad on Vermont.

Dispatch: Okay on an engine and squad at 2159.

Division 2 Attack to Command: We have the fire knocked down on Division 2. It's in the walls on Side 2 and above us.

Command: Okay. I'm changing your designation to Division 2 Overhaul. Open up.

Engine 17: Okay on Overhaul on Division 2.

Dispatch: Okay on Engine 17 from Division 2 Attack to Division 2 Overhaul at 2202.

Division 2 Overhaul to Command: We have fire all over up here. It's in the walls. We need another crew with a line and pike poles.

Command: Okay. When my next unit arrives, I'll send them up to assist you.

Ventilation to Command: Ventilation complete. One large hole. Good heat coming out. The roof is beginning to feel spongy. We're coming down.

Command: Okay. When you get down, position your ground ladder to one of the front windows of Division 2 on Side 1.

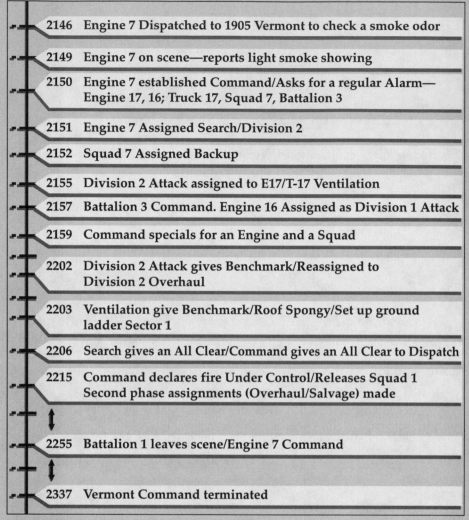

Time	Event
2146	Engine 7 Dispatched to 1905 Vermont to check a smoke odor
2149	Engine 7 on scene—reports light smoke showing
2150	Engine 7 established Command/Asks for a regular Alarm—Engine 17, 16; Truck 17, Squad 7, Battalion 3
2151	Engine 7 Assigned Search/Division 2
2152	Squad 7 Assigned Backup
2155	Division 2 Attack assigned to E17/T-17 Ventilation
2157	Battalion 3 Command. Engine 16 Assigned as Division 1 Attack
2159	Command specials for an Engine and a Squad
2202	Division 2 Attack gives Benchmark/Reassigned to Division 2 Overhaul
2203	Ventilation give Benchmark/Roof Spongy/Set up ground ladder Sector 1
2206	Search gives an All Clear/Command gives an All Clear to Dispatch
2215	Command declares fire Under Control/Releases Squad 1 Second phase assignments (Overhaul/Salvage) made
2255	Battalion 1 leaves scene/Engine 7 Command
2337	Vermont Command terminated

Figure 24-23. Incident time line.

Command has now provided a second means of egress from the second floor if working crews on Division 2 need it.

 Command to all units on Division 2: There is a ground ladder up on Division 2, Side 1, Sector A. (Every officer can close his eyes and envision second floor, front left side of the building.)

Dispatch: Okay on Ventilation complete and ground ladder to Side 1 at 2203.

Command gets an "All clear" from Search. He tells Search to come down and take a break. They've done enough for a while.

Scenario #4

Incident Worksheet

Address: *1905 VERMONT*	Owner: *WM. A. Johnson* Address: *1754 STONE CT.* Phone: *555-1114*	Occupant: *BENS CONVENIENCE STORE* Phone: *BEN WILSON* *1943 VERMONT — HOME*
Detectors: Yes (No) Worked		*PN — STORE 555-6341* *HOME — 555-2701*
Cause:	Utilities Off—Gas Elec Water	

Functional Sectors:

		Vehicular Accident
Fire ✓	*Haz-Mat*	Patient Care
Attack ✓	Entry	Extrication
Search ✓	Decon	Backup
Backup ✓	H.M. Safety	Traffic Control
Ventilation ✓	Perimeter	Transport
~~Exposures~~	Medical/Rehab	
~~Extension~~	Evacuation	
Overhaul ✓	Backup	
Salvage ✓	Scene Safety	

Geographical Sectors:

Division=Floor
Base=2 floors below fire

(sketch of building sectors: 3/3 top, 2/2 left, 4/4 right, B C / A D center, 1 C.P. bottom)

Fire Protection System

Sprinkler *NO*
Standpipe _____
Other _____
Connection Location _____
Risers _____
Fire Pump Cap. _____
Misc. _____

Construction Type

Cons. Type *BALLOON*
Roof Type *FLAT TAR*
Truss: _____
Occupancy *STORE FRONT*
~~Elevators~~ *W/ FOUR APTS*
Misc. _____

Hazardous Materials

Material _____
I.D.# _____
L.E.L. _____
U.E.L. _____
Specific Grav. _____
Vapor Dens. _____
T.L.V. _____ ppm
Readings _____

Sketch:

Unit	Assgn# 1	Bench-mark	Assgn# 2	Bench-mark
E7	SEARCH DIV2	—	REHAB	
SQ7	BACKUP	/		
E17	DIV2 ATTACK		DIV2 OVERHAUL	DIV3 SALVAGE
T7	VENT			
E16	DIV1 ATT		DIV1 OVERHAUL	DIV1 SALVAGE
E5	ASSIST DIV2 OVERHAUL			
REHAB	— MEDIC 3			

Figure 24-24. Tactical worksheet.

Command to Dispatch: I have an "All clear."

Dispatch: Okay on "All clear" at 2206.

Engine 5 reports that it is staged. Command assigns it to report to Division 2 Overhaul and to bring up a line off Engine 16 and some pike poles. About 10 more minutes pass. It looks pretty good now from the outside. Command still has many things that need to be done, but the tone is a bit more relaxed now. He has several key benchmarks from sector officers. He knows that five people were assisted out of their apartments by search and that everyone is now accounted for. All victims have been checked by the medic unit. No one required transport. The fire is knocked down on both divisions. The roof is vented. A backup line is on the rear stairs.

 Vermont Command to Dispatch: This fire is under control. Squad 1 can go back in service.

Dispatch: Okay on "Under control" and Squad 1 in service at 2115.

WHAT TO DO NEXT

- Change Division 1 Attack to Division 1 Overhaul. Remind them to check for the area of origin (focus).
- Send all units through rehab when they come out. This has been a tough fire. It's very hot and humid out.
- Check on the displaced victims. Do they need housing, clothing, and so on?
- Get an investigator on the scene.
- Start salvage operations on Divisions 1 and 2.
- Tell everyone they did a good job at the tailboard critique.
- Get the crews back as soon as possible.

Command reassigns the initial companies to the second-phase sectors of Overhaul and Salvage. This happens in the time frame between 2115 and 2155. All crews are put through rehab. After rehab, they are assigned functional sectors, usually relieving crews who then will report to rehab for their break.

Battalion 1 passes command back to Engine 7 and returns to service at 2155. After final pickup and structure checks, Vermont Command is terminated at 2237.

COMMAND STRUCTURE

Initially, Engine 7 was the only unit on the scene. One the officer of Engine 7 established formal command, he assigned his crew to search on Division 2. Figure 24-20 indicates this very simple structure. Prior to the arrival of Battalion 3, the Engine 7 officer had made three more assignments. Note that the span of control was not stretched (see Figure 24-21). When Battalion 3 took command, a significant command structure was in place. The previous incident commander knew where every unit was on the scene and its task or focus (assignment). Battalion 3 Command assigns one more sector (Division 1 Attack). As previously indicated, Command still has several functional assignments to make. He special-called for additional units, one of which was assigned to assist a previously assigned sector. The other unit was staged and sent back without being used. Because of this, Command still did not exceed the effective span of control. He used on-scene crews to handle these reassignments.

Command was passed back down to Engine 7, and the incident was closed at 2237.

INCIDENT TIME LINE

This incident had a rough and confusing start. One officer arrived at an incident that was very dynamic. Command recognized this and immediately summoned more help. Once units started to arrive on the scene, appropriate assignments were made in a logical, timely manner. All of the "things" that would cause concern (negligence) at this incident were covered by the transmission between Command and Dispatch.

Endnotes

1. I refer to these fires as "chasers" because you end up chasing the fire all over the structure. If the fire gets a good foothold on the structure, you will find fire almost everywhere you sink an ax.

DISCUSSION QUESTIONS

Note: Since this entire chapter is a review and application of the information presented in previous chapters, only discussion questions are presented. Such discussions enhance understanding of the underlying principles and how they can best be applied to an individual department.

Case Study 1

1. Review Command's actions. Do you concur with his initial sector assignments, or would you have done things differently? If so, give examples.

2. How do you feel about fighting this fire with only booster-tank water?

3. Discuss the activities concerning backup.

Case Study 2

1. Review Command's actions. Do you concur with his initial sector assignments, or would you have done things differently? If so, give examples.

2. Discuss the positioning of the backup line. Could it have been positioned in a better location? If so, why?

Case Study 3

1. Review Command's actions. Do you concur with his initial sector assignments, or would you have done things differently? If so, give examples.

2. Review Command's to-do list. Would your list have the same items? If so, would they be in the same order?

Case Study 4

1. Review Command's actions. Do you concur with his initial sector assignments, or would you have done things differently? If so, give examples.

2. Would you have given Command the same two options listed in the chapter? If you were Command, what assignment would you have made first?

3. The author presents a list of the tasks that remain to be done on page 404. With the units on the scene and in light of their last known assignment, make the last assignments necessary at this fire (there may be more than those listed in the text).

Appendix A

*T*he following sample Incident Command System procedure is provided for your review. This sample meets NFPA 1561, Incident Management System, and the Consortium criteria. The sample sector designations were left blank intentionally. Fill in your own terminology. Unfortunately, there still is no one nationally recognized method of designating areas in and around fire buildings and accident scenes.

SAMPLE INCIDENT COMMAND SYSTEM

The _____ Fire Department shall use the Incident Command System at every incident to which it responds. All members shall review and understand the following procedure:

I. Command. We shall use the Incident Command System at every incident to which we respond. The first unit on the scene of every incident shall establish command. The Incident Command is directly responsible for the outcome of the incident. Consequently, no other firefighting unit on the scene or responding, regardless of rank, shall give direction to or take direction from anyone other than the Incident Commander. (The exception to this would be if Command designates an Operations sector.)

Three types of command will be used, depending on the number of units responding and the number of entities represented at the command post. The three types of Command are as follows:

a) Informal: This type of command is used when only one unit responds to an incident. When a single unit responds to an incident, it does not formally have to announce that the officer is Command. Being the only officer on the scene, it is understood that only one officer is on the scene and that he or she is in charge.

b) Formal: This type of command shall be established whenever more than one unit responds to an incident. The first unit on the scene shall announce that he is Command along with the normal on-scene announcement. From the time that Command is formally established, no one shall do anything at the incident unless directed to by Command.

c) Unified: This type of command shall be used at large incidents when the incident would be best served if more than one entity shares the responsibility of command.

As an incident expands, Command must gather the necessary resources to handle the situation. Command must ensure that a reasonable span of control is maintained throughout the incident. Normally, one officer shall not be responsible for more than five (5) subordinates at any one time. This one-to-five span of control ratio transcends all levels of the command structure at all incidents.

Command shall be responsible for the following four functions at every incident:

a) Operations: Directly mitigates the emergency or situation present. At a fire, Operations directs firefighting crews. At a haz-mat incident, Operations ensures that the steps needed to stop the release and secure the area are taken. All emergency response crews operating at an incident report to Operations.

b) Planning: Responsible for looking at what has happened, what is currently happening, and what will (or can) happen in the future concerning the incident. Planning considers and records specifics concerning the incident itself and the crews operating at the incident.

c) Logistics: Responsible for getting the tools and equipment necessary to handle the situation. Once specific tools and equipment are on the scene, Logistics is responsible for their maintenance and replacement when needed.

d) Administration/Finance: Handles the administrative needs of the incident such as the payroll, payments for purchased equipment, and the handling of workers' compensation claims.

Additionally, Command has the following responsibilities pertaining to staff at every incident:

1. Safety: the individual responsible for the safety concerns of all on-scene personnel and civilians in close proximity to the incident.

2. Liaison: the individual who must interact with the outside agencies that respond, report to, or are present at emergencies, including the police and other law enforcement agencies, the public utilities, local and other government officials, and agencies and civilians who have a vested interest in the incident.

3. Information Officer (or PIO): the individual responsible for providing information concerning the incident to the news media, concerned outside agencies, displaced and/or affected civilians at the scene, and safety crews.

If Command cannot handle all of the above-mentioned functions and staff positions, he shall delegate any or all of them to other members at the scene. If Command does not delegate these responsibilities to others, then Command is responsible for them. If, due to the scope and needs of the incident, Command assigns staff sections, those members shall manage the responsibilities of that specific function, bearing in mind that Command must be informed and kept up to date with regard to developments and the actions of each specific function.

The first-arriving unit on the scene of the emergency shall establish command. If the first unit to arrive has more than one member, the officer (if one is riding the unit) shall become the Incident Commander. If the first member to arrive has only members of like rank, the _____ (most senior—example) member shall take command of the incident. Once someone formally takes command of an incident, all other responding units shall stage at the closest intersection to the incident in the direction of their normal response until given directions by Command. The member establishing command will remain as Command until the incident has been terminated or until command has been transferred to a higher ranking officer (usually the responding chief officer). If command is passed up the chain of command during the incident, it shall be passed down the chain upon de-escalation of the incident, ensuring that someone from the department will always be in command of every incident. Command shall make such assignments and assemble a command structure proprotional to the needs of the incident and the resources available.

II. Staging. Staging is the placement of personnel and equipment that are ready for immediate use. Staging gives unassigned responding units focus and a place for which to look.

There are two levels of staging. Level I is used for regular or first-alarm (or less) assignments. The officer of the unit responding chooses the appropriate location to stage. Engines should stage at or near a water source. Truck companies should stage at an intersection that provides access to the front of the building. Other units responding (except for chief officers) can stage at the intersection closest to the incident in their direction of response. Chief officers can report to the incident scene or stage at an intersection. It's their choice.

Level II staging requires that responding units report to a specified staging area. Command usually will designate the staging area. Level II staging normally is used at large, more complex incidents such as second alarms. If Level II staging is established, all responding units shall report to the designated staging area, not the incident. The first unit (officer) who reports to the staging area shall be designated the Staging Officer. The staging officer (normally call "Staging") is responsible for all activities at the staging area and for sending to the incident scene the appropriate units requested by Command or Operations.

III. Sectorization. Sectorization is dividing the incident into manageable units. There are three types of sectors.

a) Functional: defines a specific task. Functional sectors are referred to as _____ (such as Attack _____).

b) Geographic: defines an area. Geographic sectors are referred to as "_____" (such as West _____ or _____ C). Floor levels are called _____.

c) Combination: may be functional and geographic. An example would be _____ Attack or _____ Search. The following diagrams indicate how we will sector incidents.

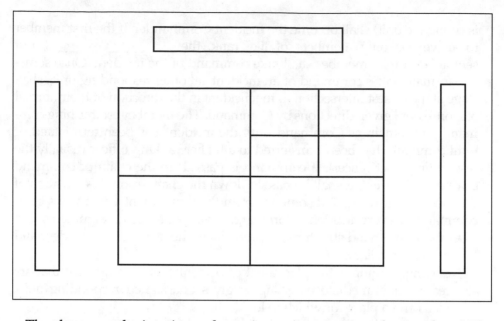

The above are the interior and exterior assignments used at incidents. The diagram below indicates floor levels.

IV. Other Incident Command Procedures:

a) Command as established and followed above will apply to emergency medical incidents also.

b) Once functional or geographic sector assignments have been made, a company or unit will be referred to by its assignment name instead of its unit assignment. (If Engine 2 is assigned to be Attack, its new radio designation will be "Attack.")

c) A system will be implemented to track the names and locations of all members on the scene at every incident. We will use the _____ method of personnel accountability. This is further explained in Section _____ of our _____ procedures manual.

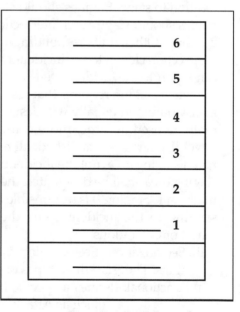

d) A system for rest and rehabiliation (rehab) shall be used at incidents that continue for longer than _____ or whenever an individual firefighter uses _____ SCBA bottles at any incident. This rehab system is further explained in Section _____ of our _____ procedures manual.

The following flowcharts depict the use of the Incident Command System.

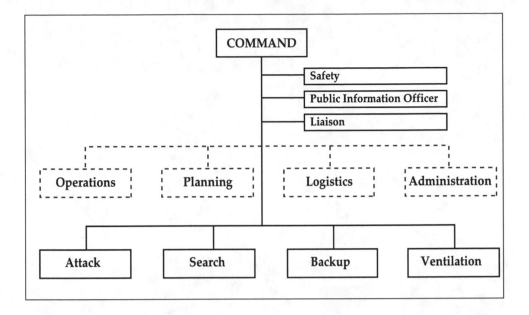

This flowchart depicts a small incident such as a single-family residential structure fire. Sections in dotted boxes or with dotted lines are those that Command has not passed off to others. In this situation, Command has assigned Attack, Search, Backup, and Ventilation sectors.

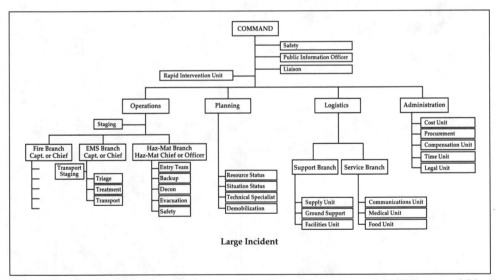

This flowchart depicts a large incident. Just about everything that could be assigned under Operations and the Staff sections is indicated. The units under the Fire Branch were intentionally left blank so that your own terminology can be inserted.

Appendix B

COMMAND & UNIT ASSIGNMENT

UNITS		
ENGINE COMPANIES		
CO #	ASSIGNMENT	SEC.
TRUCK COMPANIES		
CO #	ASSIGNMENT	SEC.
OTHER		
CO #	ASSIGNMENT	SEC.

I.C.
COMMAND POST LOCATION
SAFETY OFFICER
PUBLIC INFORMATION OFFICER
LIAISON:
OPERATIONS: ATTACK GROUP————————   RESCUE GROUP————————   EXPOSURES GROUP————————   VENTILATION GROUP————————   STAGING OFFICER ————————   LOCATION ————————   EXTENSION GROUP————————   SECTOR ————————   LOCATION————————   SECTOR————————   LOCATION————————   SECTOR————————   LOCATION————————
PLANNING: RESOURCES ————————   SIT STAT————————   DOCUMENTATION ————————   TECHNICAL ————————
LOGISTICS: COMMUNICATIONS OFFICER ———   EMS OFFICER ————————   FOOD UNIT————————   SUPPLIES————————   FACILITIES————————   SUPPORT GROUPS ————————
FINANCE ————————

INCIDENT WORKSHEET STRUCTURE FIRE

Address:					
Owner:		Address:			Ph:
Occupant:					
Structure Type:		# Stories:		Detectors:	

Unit	Assgn 1	Assgn 2	Benchmark	Other	Yes	No
				F.I.U.		
				Gas Co.		
				Electric Co.		
				Sprinklers		
				Standpipe		
				Police #		
				Air Unit		
				Safety Off.		

INCIDENT WORKSHEET

Address:			
Owner:		Address:	Ph:
Occupant:			Ph:
Structure Type:	# Stories:		Detectors:

Unit	Assgn 1	Assgn 2	Benchmark

Other	Yes	No
F.I.U.		
Gas Co.		
Electric Co.		
Sprinklers		
Standpipes		
Red Cross		
Police #		
Air Unit		
Safety Off.		

Reminder Sheet

Initial Incident Considerations

Command Type			Command Post Type	
Units Dispatched	Command Type		Type of Command	Command Post Type
Only One	Informal		Informal	Mobile
More Than One	Formal		Formal	Stationary or Formal
			Unified	Formal

Command's Priorities

Firefighter Safety ☞	Civilian Safety ☞	Stop the Problem ☞	Conserve Property

Initial Assignments

"Big Four": Fire	"Big Four": EMS	"Big Four": Haz-Mat
Attack	Firefighter(s) needs—Gloved-up, scene safe	Decon
Search	Victim(s) needs—Patient Care	Entry
Backup	Transport	Haz-Mat Safety
Ventilation	Scene Security	Backup

Benchmarks

Command - Fire under control.	Exposure -Exposures covered.	Triage -Victims triaged.
Attack -Fire knocked down.	Extension-Extensions checked.	Treatment -Victims treated.
Backup -Backup line in place.	Overhaul -Overhaul complete.	Transported-Victims transported.
Search -All clear or complete.	Salvage -Salvage complete.	Extrication -Victims extricated
Ventilation-Ventilation complete.	RIT -Members located/outside.	Haz-mat -Leak stopped/spill controlled.

Fire Attack Types

Offensive Attack: Quick Interior		Defensive Attack: Exterior	
Direct	Straight stream at seat of fire	Direct	Straight stream at flame
Indirect	Straight stream off ceiling	Indirect	Used for backdraft situation
Fog	Fog pattern—Holding action	Combination	Fog swirled from outside window, fire in flashover

(Over)

Flowcharts and Assignments

Write unit or names of members filling the roles below. Boxes with dotted lines indicate functions that Command has retained. If Command passes off a section with a dotted box, write in the name of the member assigned. Not every box and line need be filled at every incident. Use these charts to help track units and assignments at incidents.

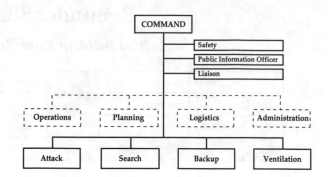

First Alarm Structure Fire

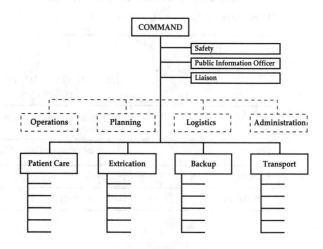

Vehicular Accident

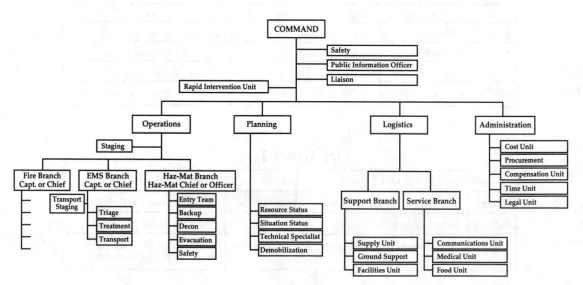

Large Incident

Answers to Chapter Review Questions

Chapter 1

1. In the late 1960s and early 1970s, California was plagued by a rash of wildland fires. After a particularly large wildland fire in California in 1970, a group of chiefs and state officials determined there had to be a better method of handling such large fires.

2. Federal funding was provided to create a standard system that would be used at these incidents.

3. *FIRESCOPE*; the state forestry industry and wildland firefighters.

4. The National Fire Academy's (also called *NIIMS* and *FIRESCOPE*) and the Phoenix Method.

5. The responsibilities of the first-arriving officer, the responsibilities of Command, and the manner in which the incident is divided.

Chapter 2

1. Standard, all.

2. Roles and relationships.

3. Freelancing.

4. One person shall be in charge of every incident to which we respond. Command ultimately is responsible for the outcome of the incident. If Command is to be held ultimately responsible for the incident, he must have total control of the incident.

5. The person ultimately responsible for the outcome of the incident.

6. That one—and only one—individual is in charge; everyone responding (or who could respond, as in mutual aid) knows what the individual in charge is called; and no one does anything until that person says to.

7. SARA Title III.

8. The number of units responding to the incident, the number of entities formally represented at the command post.

9. Informal, formal, unified.

10. Informal command: only one officer/unit responds and only one entity is at the command post; formal command: more than one officer/unit respond, only one entity is at the command post; unified command: more than one officer/unit respond, more than one entity is at the command post.

11. Command, staging, sectorization.

12. Command, communications, span of control.

13. To focus on the whole.

14. Firefighter safety, civilian safety, stop the problem, conserve property.

15. Operations, planning, logistics, administration.

Chapter 3

1. The fact that he is on the scene, an initial conditions report, the command mode that will be used.

2. Command.

3. Are all the appropriate sectors assigned and in good working condition? Is the incident still expanding or beginning to de-escalate? Are the needs of the occupants being met? Are the needs of the fire crews being met?

4. False.

5. True.

6. The situation.

7. The on-scene report.

8. They depend on what the officer first observes on arrival.

9. Investigate the situation.

10. Stage.

11. The next officer to arrive on the scene.

12. The first unit on the scene handles the entire incident and cancels the other responding units and reverts to informal command. The first officer determines that the incident is too involved and establishes formal command. The next officer on the scene establishes formal command on his arrival.

13. The extent of the incident on arrival, the location of other responding units.

14. Nothing showing, fast attack, command.

15. Name an acting officer, assign the crew to another crew, give them staff positions.

Chapter 4

1. The area within which Command is expected to operate.

2. Command.

3. When Command leaves the command post for a short period of time.

4. To talk to the occupants, to check on the status of injured civilians or firefighters, to begin origin and cause determination, to quickly check the status of exterior fire areas.

5. At the incident site, so that the focus can be on the needs of the incident.

6. Mobile, stationary, formal.

7. The type of command established, the needs of the incident at the present time, the potential for the incident's expanding.

8. The new location provides a better view of at least two sides of the incident, the atmosphere of the original location is exposing command post occupants to smoke or other hazardous substances, the new location enables Command to move back to better focus on the whole incident, the new location is a more logical area for arriving members and outside entities.

9. Intentionally driving up to and then past the incident area to observe three sides of the involved area.

10. (Any one of the following) (a) It should be an area advantageous for focusing on the whole incident. (b) The larger the incident, the farther away the command post should be. (c) At a big incident, Command may be so remote that he cannot view the incident.

Chapter 5

1. The three predetermined roles that Command must fill at every incident.

2. Safety, liaison, public information officer.

3. Delegate them to other officers at the scene.

4. The officer of the crew.

5. Scene safety at the incident.

6. On his own or, if time permits, after consultation with Command.

7. Mobile.

8. False.

9. Firefighter safety practices, building construction, haz-mat operations, air monitoring, and confined space operations.

10. Office manager of the command post.

11. To greet persons reporting to the command post, to serve as the mouthpiece for Command, to direct outside agencies to the functions in the incident for which they are best suited.

12. Utility company representatives; the owner of the structure; representatives from local, state, and federal governments; police.

13. Interacting with the media.

14. At a location slightly remote from the command post.

15. A responding chief officer.

Chapter 6

1. Operations, planning, logistics, administration.

2. Every incident.

3. Managing the strategic aspects of the incident.

4. Making the strategic decisions concerning the incident (after consultation with Command), assigning crews to handle the situation, keeping command informed.

5. At the command post.

6. It allows operational units and Command more radio time and eliminates unnecessary disruptions.

7. To review the past actions and then address the current and future needs of the incident.

8. Any four of the following: incident planning, resource assessment, situation status, documentation, technical specialist, demobilization.

9. At the command post.

10. Getting the appropriate tools and equipment to the incident.

11. He should have a "home base" at the command post but may have to roam the scene.

12. Locate and provide requested tools and equipment, provide support for on-scene units, communications, establish a medical unit if requested, establish a food unit if requested.

13. To meet the admninstrative needs of the incident.

14. At the command post.

15. Time unit, procurement, claims unit, payroll unit, cost unit, legal unit.

Chapter 7

1. The placement of uncommitted apparatus and personnel at an incident.

2. Stops freelancing, sets a calming tone, allows you to "look important."

3. Fire department playbook.

4. Level I, Level II.

5. The level of response.

6. For incidents up to a regular alarm (or up to multiple alarm incidents).

7. The officer of the responding unit.

8. At or near a water supply.

9. If the first unit on the scene is not an engine company.

10. At an intersection that gives the best access to the front of the building.

11. At the nearest intersection.

12. At an intersection or go up to the incident; it's up to them.

13. No. The location does not change. However, if using fast attack, the second unit on the scene does not stage. The officer goes up to the incident and establishes formal command.

14. At a location advantageous for supplementing the system if needed.

15. For incidents beyond the regular alarm response.

16. Command.

17. To the staging area.

18. To control the activities of the staging area, send the appropriate units to the incident, keep Command or Operations informed.

19. What the assignment is, where the assignment is, to whom to report, any special instructions.

20. A radio, something to write on, a writing implement.

Chapter 8

1. The breaking down of the incident into manageable units.

2. Span of control, chain of command, tunnel vision.

3. If the assignment is activity- or area-based; if the activity is both activity- and area-based.

4. Functional, geographic, combination.

5. A specific activity that Command or Operations wants performed.

6. Any three of the following: attack, search, ventilation, backup, extension, exposure.

7. Areas.

8. Both activity- and area-based.

9. In the entire structure.

10. The command post.

11.

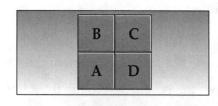

12.

13.

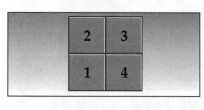

14.

15.

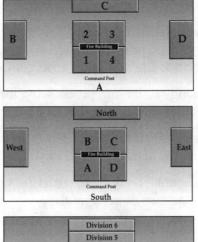

16.

17.

18.

19. Groups.

20. Sections.

21. A group of the same type of apparatus acting as a unit.

22. A group of different apparatus types acting as a unit.

23. Refers to floor level.

24. The safety of their crew, to complete the assigned task, to keep Command informed.

25. Attack, search, ventilation, backup.

Chapter 9

1. A statement that defines a specific task that a person or group of persons is sent to perform.

Chapter 10

1. To coordinate the activities of emergency crews, making every effort to use accepted strategic practices; to effectively protect life and property from the effects of fire and other emergencies.

2. The situation, using a size-up process by looking at the picture; Command's priorities; the availability of personnel and apparatus.

3. A mental process in which the picture in front of you is rapidly evaluated to determine the activities (to-do list) that are required.

4. Extent of fire and smoke on arrival, structure type, occupancy, exposure potential, the actions of occupants.

5. Confine, control, extinguish.

Chapter 11

1. To coordinate fire suppression efforts by tactically placing and directing attack lines to seek out and extinguish all fire in the area assigned and to provide feedback to Command or other assigned sectors.

2. The safety of his crew, directing his crew in attacking and extinguishing the fire in his assigned area, keeping Command informed.

3. Offensive, defensive.

4. The extent of fire on arrival, the likelihood of savable victims, the ability of crews to operate inside, the stability of the structure, the risk–benefit ratio.

5. An aggressive interior attack directed at the seat of the fire.

6. To confine, control, and extinguish the fire.

7. Direct attack, fog attack, indirect attack.

8. An exterior attack on the fire.

9. When the fire has reached flashover and at least one floor is totally involved, the structure is vacant and well-involved, fire or heavy smoke conditions exist, or there have been previous fires in the structure.

10. Holding attacks.

11. The amount of fire in the structure, the stage of burning, the stability of the structure.

12. Direct attack, indirect attack, combination attack.

13. Savable victims may be inside the structure or crews are inside the structure.

14. Let it out.

15. Toward the area of involvement.

16. Within 30 seconds.

17. Changing the floor of operations.

18. Biggest is first.

19. Start at the lowest level possible without being cut off from your means of egress.

20. Into the structure by the same entrance path taken by the company you're going to assist.

Chapter 12

1. To coordinate the primary search efforts in the area assigned.

2. The safety of his crew, to coordinate a primary search of the area assigned, to keep Command informed.

3. Primary search, secondary search.

4. A rapid systematic search of the structure to ensure that all savable victims are removed from the structure.

5. A slow, methodical search to locate deceased victims.

6. That the search officer is coordinating the search of the structure or area assigned; that Search has a specific plan in mind concerning where to search first and last; that the search officer will keep the safety of his crew paramount in his mind; when Search changes floors, he will notify Command; that Search will ask for additional help if he needs it.

7. If Command receives information pertaining to the search, he will relay that information to Search; Backup will be ensuring the safety of the search crew.

8. The time of day, the intensity of the fire and the products of combustion on arrival, the location of the fire at the time of arrival, information obtained from occupants or neighbors.

9. No.

10. No.

11. In the standard methods of search, all searchers enter the area to be searched and conduct a search by themselves or in groups. With the oriented method, one member remains outside the search area and ensures the safety of the searcher while the other member(s) conduct the actual search.

12. There must be communication between the oriented man and the searcher. Both members must know the direction of the search. The number of walls in the room must be determined.

13. The location of the door on the wall.

14. The oriented man focuses on how to get into the structure and where the search team is at all times. If the searcher and the oriented man switch positions, the searcher may not be able to ensure how to get out of the structure. It goes back to focus.

15. If more searchers are used, it becomes difficult for the oriented man to keep track of all searchers.

Chapter 13

1. To protect interior crews by pulling and strategically placing protective lines between them and obvious or potential areas of fire spread and to ensure egress if retreat is necessary.

2. The safety of all interior crews, to position and maintain a working backup line, to keep Command informed.

3. If the fire is at or near flashover and Command has committed to an offensive fire attack, multiple fires are on multiple levels of the fire building, the fire has entered and is attacking the structure, the fire is large and multiple lines are in place.

4. From the same entrance used for the attack line.

5. The number of floors in the structure, the location of the fire, whether the fire is on more than one floor.

6. Approximately two-thirds of the way between the point of entrance and the location of the attack crew.

7. The same as at a ranch house.

8. It should be taken in by the same route as the attack line—up the stairs, in an adjacent room, and then back to the hallway or landing near the stairs.

9. Normally at the top of the basement stairs.

10. Normally at the bottom of the attic stairs.

Chapter 14

1. To remove or channel the products of combustion by mechanical or natural means from the involved structure into a nonthreatening area.

2. They will rise until they exit from the structure or meet an obstruction, then they will travel horizontally until they find a way up or either accumulate or dissipate naturally.

3. To stop the damage caused by smoke and other products of combustion, assist escaping and trapped victims, make our job easier.

4. Natural, mechanical.

5. The products of combustion are removed or channeled from the structure with the aid of natural processes.

6. Convection currents, the wind, the stack effect.

7. Opening windows, putting holes in the upper levels of the structure, using the stack effect.

8. The use of mechanical devices to remove smoke and other products of combustion from a structure.

9. Fans, HVAC systems, fog streams.

10. The fire's location, the sectors assigned, the stage of the fire.

Chapter 15

1. Conduction, convection, radiation, direct flame impingement.

2. Interior, exterior.

3. Horizontal, vertical.

4. The availability of additional companies, wind direction, wind speed, the proximity of the exposed buildings, life safety, economic value, occupancy type.

5. Find the seat of the fire and then place a hose stream between the fire and the horizontal exposed areas.

6. Find the source of the fire, and use the offensive indirect attack on the fire. This not only cools the vertical exposed areas but is very effective on the source fire.

7. Choose a hose stream of adequate gpm. Direct the stream at the upper regions of the exposed building on the exposed side and let the water wash down the face of the structure.

8. The stream is directed toward the source fire to a degree that causes the exposed building to be neglected. A large enough hoseline is not used at the beginning of the operation.

9. Convection.

10. Radiation.

Chapter 16

1. To check the areas above, surrounding, and below the main body of fire for extension and to report his findings back to Command.

2. Attack.

3. Above the fire, around the fire, below the fire.

4. Sight, sound, feel.

5. Fire axes, pike poles, closet hooks.

6. Location of the fire, wind direction and speed, construction type.

7. B: assign attack sector.

8. Backup.

9. The safety of his crew; to check the areas above, around, and below the fire for extension; to keep Command informed.

10. Reassign Extension to attack in that area, take the extension crew out and assign another company to attack in that area, or have the extension crew come out and have the attack sector extinguish the fire.

Chapter 17

1. To ensure that the fire is completely out and to pinpoint the area of origin.

2. Determine the origin of the fire, ensure that the fire is completely out.

3. As soon as the main body of fire is knocked down.

4. Above, around, and below the fire.

5. Determine the area of origin.

6. **Ordinary**—Normally little if any overhaul on solid ordinary wall. If veneer wall, determine what is behind the veneer and then overhaul accordingly. **Lath and Plaster**—If you can put your hand on the wall long enough to say "It's not hot, Chief," chances are there is no fire in the wall assembly. If you must open up, use a pike pole or an ax. If you are not sure, remove a piece of baseboard and poke a small hole in that area. **Drywall on Wood Stud**—The fire load is low, but fire can enter the area. The hand-on-the-wall trick does not work because drywall holds heat. Open up with an ax or pike pole where needed. If you are not sure, remove the wall plug cover or the light switch cover. **Drywall on Steel Stud**—Little fire load. Treat as drywall on wood stud, but watch for electrical shock. Steel will conduct heat.

7. Move the debris pile away from any structure.

8. Start in an area with no damage and work back toward the area of the most damage. Check door frames and moldings. The areas with the heaviest char indicate where the fire came from. Once the room of origin has been determined, look for the areas of heaviest damage in the room and for "V" patterns.

9. No.

10. The overhaul sector works with a hoseline and puts out hot spots. The extension sector has no line. It only looks for fire. It doesn't put it out.

11. To keep the crew safe, to determine the area of origin and make sure the fire is completely out, to keep Command informed.

12. To protect as much material within the structure as practical from the effects of the fire, its by-products, and suppression tactics.

13. Ventilation, attack, extension, overhaul.

14. Covering furniture and other valuables, separating areas from the fire and its effects, protecting valuables, reminding sector officers to avoid causing unnecessary damage.

15. As soon as lifesaving and fire-control efforts are in place.

16. Where savable property is expected to be.

17. The truck company.

18. Water damage is one of the two leading causes of damage at fires. Use as little water as necessary to extinguish the fire. Don't hit smoke.

19. Close the doors to unaffected rooms.

20. To keep his crew safe, to protect savable things from the effects of the fire, to keep Command informed.

Chapter 18

1. To search for and remove trapped or injured firefighters.

2. Savable people or things are inside a structure, crews working inside are at a more-than-usual risk of injury or entrapment.

3. NFPA 1500.

4. They must be trained in the concept of RIT.

5. Fires in multifamily residential structures where rescues will be required, advanced fires in commercial structures, shipboard fires, advanced fires in high-rise structures.

6. Only if a firefighter or a fire crew becomes trapped or lost.

7. An extra SCBA bottle per member, door chocks, hand lights, rescue rope and the necessary tools associated with it, a hauling line, at least one pry bar.

8. The locations and arrangement of windows, exterior doors, stairs, fire escapes, construction type.

9. Normally at the command post.

10. All but attack crews and backup sectors in the area should leave the structure. Crews left inside should maintain their focus on their original assignments.

Chapter 19

1. Announcements that a particular activity or assignment has been completed.

2. Let Command know that an activity has been completed, lend an air of closure to the assignment, give Command a better understanding of the progress being made.

3. They should be brief, uniform, and used all the time.

4. "All clear" and "Under control."

5. In the incident time line.

6. a) j
   b) i
   c) f
   d) b
   e) c
   f) h
   g) e
   h) d
   i) g
   j) a

Chapter 20

1. The scene, the victim.

2. Meeting the needs of the crew, meeting the needs of the victim, transporting the patient, securing the scene.

3. Infection control or protecting them from acts of violence.

4. Yes.

5. To stage and await direction from Command.

Chapter 21

1. That a fire department unit has arrived on the scene, that you are at the proper location, a conditions report, that some type of command has been established.

2. That the location provides the best access to the accident location and allows other responding units access to the accident area.

3. Level II staging at a fire.

4. A crew member of the first unit on the scene.

5. Functional, geographic, combination.

6. Patient Care in the green Ford, extrication in the red Chevy.

7. Patient Care.

8. Patient assessment, immobilization/treatment, packaging, transport.

9. 1½-inch line.

10. Yes.

11. Manual, mechanical.

12. Law enforcement officials.

13. They further define responsibility and expectations.

14. Safety of the crew, to complete the task assigned, to keep Command informed.

15. Traffic, flash fires, bloodborne pathogens.

Chapter 22

1. In the case of fire and haz-mat incidents, it is the local fire department—whether it is a federal, state, city (municipal), county, township, or private department.

2. Level I, Level II, Level III.

3. Identify substance, make any feasible rescues, start evacuation.

4. Identification of the substance involved.

5. Haz-mat safety officer, entry team, backup, decon, medical unit.

6. Fire control, haz-mat, EMS.

7. Fire control, search, evacuation, perimeter.

8. Traffic control, evacuation.

9. Planning.

10. Just outside of the Warm Zone, in the Cold Zone, and slightly remote from the haz-mat unit.

Chapter 23

1. An increase in residential central station fire alarm systems and carbon monozide detectors.

2. Firefighter safety, civilian safety.

3. Haz-mat incident.

4. Where the bomb has not exploded, where the bomb has exploded.

5. The police or other law enforcement agency.

6. The fire department.

7. At least a block away.

8. Under Planning.

9. Manual, mechanical.

10. Unified.

Index